FIRE!

FIRE!

THE 100 MOST DEVASTATING FIRES
AND THE HEROES WHO FOUGHT THEM

By Edward C. Goodman

BLACK DOG
& LEVENTHAL PUBLISHERS
NEW YORK

ISBN: 1-57912-160-8

Library of Congress Cataloging-in-Publication Data

Goodman, Edward C., 1952-
  Fire! : the 100 most devastating fires through the ages and the
heroes who fought them / by Edward C. Goodman.
        p.   cm.
    ISBN 1-57912-160-8
    1. Fires. 2. Explosions. 3. Fire extinction. I.Title.
  TH9448 .G66 2001
  363.37—dc21   200102598

Published by
Black Dog & Leventhal Publishers, Inc.
151 West 19th Street
New York, N.Y. 10011

Distributed by
Workman Publishing Company
708 Broadway
New York, N.Y. 10003

h g f e d c b a

Printed in Spain
D.L. to: 779-2001

Acknowledgments

Many people helped shape this book. I am most indebted to my editor, William Kiester, who had a vision of the book and provided critical suggestions, which substantially reshaped the book.

I am grateful to everyone at Black Dog & Leventhal, particularly Pam Horn; Anne Burns the photo researcher; Dutton and Sherman, the design firm; and especially J.P. Leventhal who was enthusiastic about the project from the beginning and gave me the opportunity to create it.

I want to thank my parents Harry K. and Shirley E. Goodman and my sister Helen J. Goodman for their love and continued faith in me.

I am indebted to my college friends Bill Godfrey, Edward J. Donovan and Jerome D'Angelo with whom I tossed around the initial ideas for the book while on vacation.

Special thanks go to Judith Dupré who inspired me and encouraged me to write and Tim Stauffer who introduced me to J.P. Leventhal.

Many other people have contributed their expertise and resources to me. I am indebted to the following persons for their contributions and support:

Angela Giral, Claudia Funke, and Janet Parks, Avery Architectural and Fine Arts Library; Mark Vivian, Mary Evans Picture Library; Joan A. Kissell; Kathleen Kerr-Curran; Patricia, James, Madeline and Kathleen Carver; Lesley Bruynesteyn, my former editor.

Contents

Introduction

Fire! Mystical and magical. Tamed and wild. The very word frightens us with its destructive power and yet awes us with its strength.

Throughout history, fire has fascinated mankind. Myths of the origin of fire have been seen in every society. From Australia to South America to ancient Europe, the tales of birds or animals bringing fire to man abound. The legend of Prometheus stealing fire from the gods to give it to mankind is one of the most famous Greek legends. Aristotle declared that fire was one of the four elements of all things along with water, earth and air. Fire was also used in religious ceremonies. In the ancient Vedic scriptures, *Agni*, which means "fire," was the messenger between the people and the gods. In ancient Rome the vestal virgins guarded a perpetually burning sacred fire. Among the Hebrews fire was seen as both a punishment and a powerful sign from God.

The burning of ancient cities such as Troy have become legends. Who has not heard of Nero fiddling during the six days in July 64 A.D. when the city was reduced to ashes? He actually played the lyre and sang a song of his own composing "The Burning of Troy."

Fire has transformed the domestic and commercial development of our society. Early man used fire as a means of warmth and a way to cook food. Fire was tamed and used to clear land and expand agriculture. The Industrial Revolution harnessed fire as a fuel to power expanding industries, which in turn increased the size of our cities and brought industrialized society to every corner of the globe. Fire has been the basis for transportation, industry and agriculture. In the 19th century it was said that everyone used fire. It lit the lamps at night for reading, powered the kitchen stoves, and warmed the office buildings and schools. It is the image of hearth-and-home that is fundamental to the idealization of family.

And yet fire is also a destructive force. We are both terrorized and enthralled by stories of fires. The fear of a painful death by burning reaches across history from martyrs being burned at the stake to more contemporary tales of people leaping to their deaths to escape the terrible pain of death by fire. We are enthralled by stories of survivors but fearful of being caught in a fire ourselves. Early accounts of fires, particularly in the 19th century yellow press, were sensationalistic and particularly gruesome in the description of the victims. Phrases such as "heaps of half-charred, twisted bodies" were a common motif in newspaper accounts at the time.

The devastating fires that burned cities such as Chicago, Baltimore and San Francisco stemmed to a large degree from the use of non-fireproof building materials such as wood as well as overcrowded living and working conditions. The growth of cities and the need to protect homes and businesses from the perils of fire is seen in the rise of firefighting.

The history of fires is wrapped up with the history of the efforts to fight fires. From ancient times to the present day, fire fighting relied on individuals uniting as a team to fight the flames. Nomadic man could do little to fight the wildfires caused by lightning, but as hunters and gatherers settled into farming and formed villages, firefighting became more important. In 1st century B.C. Alexandria, enormous water-filled syringes were used to fight fires. Volunteers in medieval Europe used wooden and leather buckets to transport water and hooks on poles to pull down buildings to create firebreaks. By the time of the Great Fire of London in 1666, wagons filled with water were equipped with hand-operated pumps that could shoot water a short distance, although they were no match for a fire of any size.

Colonial cities in America used volunteer fire watchmen who walked through the towns at night checking for fires. The first permanent volunteer fire company, the *Union Fire Company*, was founded by Benjamin Franklin in Philadelphia in 1736 followed by New York City in 1737. However, rival volunteer fire companies often fought with each other at the scene rather than extinguish the flames.

The first steam-powered fire engine was developed in London in 1829 and it could throw a stream of water over 80 feet. At first teams of men were used to haul the early pumpers; horses were employed the 1850s. Ladders to help people escape had been mounted on wagons in the 1700s and about 1832, sectional ladders, which could be joined together, were being used in London. In 1837 the first portable, extension ladder mounted on wheels was invented and could reach a height of 65 feet.

The devastating fires in 19th century American cities like Chicago, Boston and Seattle and individual European fires like the Ring Theater in Vienna were blamed on poor construction materials that were not fireproof. Insufficient water supplies to fighting fires were also cited. There were so many large fires in American cities, that the nearly bankrupt insurance industry formed the National Board of Fire Underwriters (NBFU), which began to evaluate the level of fire-preparedness of departments in cities.

By 1900 most large cities had paid fire departments, steam pumps mounted on gasoline-powered trucks, fire alarm systems and fire hydrants attached to water systems.

City fire departments were unavailable in rural areas where wildfires frequently raged out of control. Wildfires begun by lightning strikes provided the first source of fire for mankind and through the centuries, wildfires have burned farms, cabins and entire towns. Today, lightning strikes account for over 10 percent of the wildfires each year. Man is responsible for the remainder both through arson incidents as well as "controlled burns" set by local fire officials to clear brush that sometimes grow out of control and can be difficult to extinguish. This is how the massive Los Alamos fires in New Mexico started in 2000.

Nearly all the advances in fire-fighting techniques we see today have been created this century, including requirements for adequate exits, sprinkler systems, revised building codes and fireproof materials. Postwar technology has allowed aerial firefighting using airplanes and helicopters, and improvements in off-road vehicles have given ground-based crews a greater variety of equipment such as bulldozers and four-wheel drive fire engines.

Modern firefighters worldwide are now skilled in using a variety of equipment and respond to a wide variety of emergency situations from explosions and terrorism to hazardous material leaks.

There are more than 30,000 fire departments and over a million firefighters in the U.S. The job of the firefighter is still dangerous. Every year in America, there are approximately 100 firefighters killed with more than 100,000 injured. These dedicated heroes save thousands of lives and millions of dollars in property each year.

This book is not intended to be comprehensive but is a selective survey of some of the most destructive international fires through time. It is organized chronologically and includes the progress of each fire, the stories of the victims and firefighters and the results of the fire.

With more than 7 million fires annually worldwide, fire continues to be both a friend and foe to mankind. It has allowed us to achieve the highest degree of civilization ever seen and yet can reduce our achievements to charred ruins in moments.

Moore Fields

Spittle Fields

...reof since the
...dent of fire
...space signifeie
& where the h
prest, those pl

Sould by Iohn Overton at
littie Brittaine, next th
gak 1666

RIVER THAMES

The Tow

London, England

DATE · SEPTEMBER 2-6, 1666
DAMAGE · £10,000,000
CAUSE · BAKERY ACCIDENT

Samuel Pepys tells readers of his diary that at 3 a.m. on Sunday morning, September 2, one of his maidservants awakened the household to tell them of a fire seen in the city. "So I rose," he writes "and went to her window…but being unused to such fires as followed, I thought it far enough off and so went to bed again, and to sleep."

The Great Fire began in the King's baker's house at 25 Pudding Lane, near London Bridge. The baker later testified that three hours after he had gone to bed he awoke and found flames licking the woodwork of his staircase. A strong east wind carried sparks across the narrow lane to an inn and from there to the wharves where hay, coal, oil and timber were stored. From a minor incident it became a major danger.

The fire-fighting equipment of the period consisted of water buckets, water hand-squirters, which were similar to big turkey basters, and long fire-hooks on poles for pulling down buildings to create fire breaks. If the Lord Mayor had ordered these tools to be used, the fire might have been contained but, fearing claims for reimbursement for destroyed buildings, he hesitated until it was too late.

For the next three days there was pandemonium. Wild rumors spread that the fire was due to an invading force of French or Dutch or Papists. Pepys rode out on the

---

ABOVE *The 1666 London fire destroyed ⁵⁄₆ of the city, as seen in this map drawn up just after the fire. The white part of the map (without the buildings drawn in) is the burned area.*

The City Before the Fire

Medieval London was a walled city of narrow streets and timbered houses, which had expanded westward. Within the walls stood Old Saint Paul's church and a hundred other medieval churches, along with taverns, banking houses, businesses and warehouses, mostly made of wood.

ABOVE *People, carrying their belongings, flee through Ludgate, one of the narrow city gates. In the background is Old St. Paul's church, which was destroyed.*

river to inspect the fire and took the news to King Charles II at Whitehall. People, laden with household goods, pushed their way through the narrow city gates, fighting with each other to escape. Carts broke down and terrified horses ran wild. All London might have burnt down, except that King Charles, seeing that the Mayor had lost control, put the Duke of York in supreme com-

mand, with members of the Privy Council under him. Fire-fighting stations were established in an arc around the fire, manned by 30 soldiers each.

Tuesday, after three days of burning, was the worst day. Fire burned St. Paul's, the Guildhall, Newgate and Fleet Street. It also marked the beginning of the end of the fire. Sailors called to duty began to blow up entire blocks of houses to form a giant firebreak at the fire's front, while the militia pulled down many more buildings on other flanks of the fire. King Charles himself worked like any laborer, handling buckets alongside his

"You see the dismal ruins the fire hath made; and nothing but a miracle of God's mercy could have preserved what is left from the same destruction."

—King Charles II

brother the Duke of York. The counter-measures began to take effect.

Wednesday, the wind died down, stopping the spread of the fire and allowing the firebreaks of torn-down buildings to hold. By midnight, the fire was almost out and finally ended at Pye Corner after burning $^5/_6$ of the city, including 87 parish churches and at least 13,000 dwellings. Thousands of refugees from the fire were camped out in makeshift tents in the fields north and west of the city. Royal proclamations provided food and essential services. Workmen began to clear debris and a week later on September 13, the King declared he would rebuild in stone or brick and widen the streets to help prevent this from happening again.

RIGHT *"The Golden Boy of Cock Lane" is located in East London, on the corner of Cock Lane and Gulpin Street, to mark the spot where the fire ended.*

BELOW *People used any means possible to escape the flames. Here resourceful refugees can be seen fleeing as the city burns in this print of Lieve Verschuier's "The Great Fire of London, 1666."*

An Exact Prospect of CHARLES TOWN.

Charleston, South Carolina

DATE · NOVEMBER 18, 1740
DAMAGE · £250,000
CAUSE · HATTER'S SHOP
ACCIDENT

Colonial cities in America were constantly endangered by fires, due to their wooden buildings and narrow streets. The "Great Fire" of 1740 in Charleston was one of the worst in colonial times.

The fire began on the afternoon of November 18, when an accident in a hatter's shop, probably a spark from a tool, ignited a small blaze. Strong winds soon whipped the fire into an inferno, which spread with great rapidity throughout the building and threw sparks on to adjacent buildings. The wooden houses and warehouses near the river fed the flames. As the blaze spread, panicked families rushed to escape the flames, loading their possessions in carts, which clogged the streets and hampered fire-fighting efforts.

"From one of the most flourishing towns in America, Charleston is at once reduced to ashes."

—Robert Pringle, survivor

The fire continued throughout the day and was only extinguished when British troops from ships in the harbor arrived and began to pull down houses to form a firebreak. The fire burned more than 300 houses and businesses in the center of town and left hundreds homeless and jobless.

ABOVE *"View of Charles Town" painted in 1739 by Bishop Roberts, one year before the great fire.*

Metropolis of the Province of SOUTH CAROLINA.

The Relief

Historian Matthew Mulcahy recounts that, "People burnt out by the fire began to seek aid as soon as the flames were extinguished. Those most destitute turned to St. Philip's Church for relief." Money was immediately distributed to the needy, but the church was soon overwhelmed with requests. Donations from individuals, such as William Bull, the Lieutenant Governor of the colony, began to arrive and continued to help those in need survive through the winter. In January, 1741, the South Carolina Assembly provided a major grant of money that was distributed to 17 families. According to Mulcahy, "Bull wrote to England and pleaded for help from Parliament, [which]...granted £20,000 sterling in

April." Boston and Philadelphia sent food and money immediately. The money from Britain was intended for rebuilding the city and a local committee of eight judged the need of applicants in 1742. Many of the people awarded money from this grant were unfortunately the least needy—businessmen and politicians. The Assembly also smartly passed building regulations, which required houses to be built of brick and set back from the street.

FIRE-FIGHTING TECHNOLOGY AT THE TIME

Charleston had no formal city government and no organized fire-fighting brigades, unlike other large colonial cities such as New York and Philadelphia. Leather buckets and hooks were supposed to be in most buildings to fight fires. The hooks were used to help pull down houses in hopes of stopping the spread of fires. They had been brought to the American Colonies from England, where they had been used in the Great Fire of London. But this rudimentary yet fairly effective equipment was not used. Rather than fighting the fire at the beginning, when it might have been contained, most residents were more concerned with saving their possessions and fleeing.

Esparanza Dutch Consuls Quinta St Catharine St Francisco La Moira

Lisbon, Portugal

DATE · NOVEMBER 1, 1755
DAMAGE · 15,000 DEAD
CAUSE · EARTHQUAKE

The Lisbon earthquake struck at about 9:30 a.m. on Saturday, November 1 and lasted ten minutes. It gave the town three shocks separated by intervals of a minute. A terrible rumbling was heard and then a devastating shock followed which lasted two minutes. Roofs, walls

---

ABOVE *This drawing shows the bustling city and port of Lisbon (the south-east prospect on the River Tagus) before the destruction caused by the great earthquake and fire of 1755.*

"Bury the dead, feed the living."
—Marquis de Pombal

and facades of churches, palaces, houses and shops crumbled and a dark cloud of dust rose above the city. A third tremor completed the destruction.

The amount of damage done by the earthquake was great, but many buildings, such as the Royal Palace, the new Opera House, completed in March, and the Patriarchal Church, might have survived if the fire had not started. Fires broke out in several parts of the shattered town. Fanned by a northeast wind, the flames quickly spread throughout the central low-lying section

The City Before the Fire

The city was still largely medieval in appearance, with markets at the edge of the Tagus River. A flat area called the Cidad Baixa contained businesses, the Royal Palace, government buildings and houses. The terrain slowly rose towards the Praça de Pombal in the northwest. There were over forty churches and about ninety convents and monasteries. There was fabulous wealth in its palaces and churches and the shops and warehouses held costly merchandise and jewels. Lisbon was a very rich city, which had lately come to depend on trade with Britain and Germany. Visiting merchants found that the city had three main attributes: wealth, the Inquisition and a belief in the worship of images.

ABOVE *This innovative fire engine for the time was built in England around 1750 and used in colonial Massachusetts. Firemen pump the water by pushing up and down on the wooden rods, propelling water out of the nozzle at the top.*

of town from the river onto the surrounding slopes. The fire led to more loss of wealth than the earthquake, destroying much of which might have been recovered from the ruins. The flames spared nothing.

The disaster was not yet complete. At about 11 a.m. seismic tidal waves 20 feet high crashed into the town and covered the dock area, the market and much of the central business district. Hundreds of people who had just survived the earthquake and fire were now drowned.

The fire was not extinguished for nearly a week after the earthquake because there was no one to fight it with the city falling into a state of anarchy. There were stories of people crawling with bleeding and broken limbs, frantic cries of trapped children, wounded animals feasting on the dead lying in the streets and aimless crowds wandering in search of a safe shelter. Fires and crashing walls blocked the roads leading to the country, and thieves and galley slaves who had escaped from their prisons and ships were roaming the streets, plundering in the ruins. Four hundred patients in the Royal Hospital were burned to death in their beds. Priests were passing back and forth hearing confessions and giving absolution.

Most of the population managed to settle in open areas to the east and west of the ruined city. Over 9,000 wooden huts and temporary houses were built over the next few months and all classes of people lived in them,

from royalty to bishops to laborers. Continual aftershocks frightened many and it became common to sleep outside. In August 1756 it was estimated there had been 500 aftershocks.

Restoring Order

The man to whom the rescue of Lisbon is credited is the Marquis de Pombal. Through well-connected marriages he had elevated himself from a country squire to virtually being Dictator of Portugal during the reign of José I. He advocated a practical approach to cleaning up following the fire. He ordered the quick burial of all dead, began to clear streets, seized all food supplies for miles, as well as whole cargoes of goods from ships in the harbor, opened food centers and controlled price gouging. He ordered the army to execute robbers, had engineers examine surviving buildings, cleared the ruins and began a grand rebuilding scheme. He took control of a very difficult and chaotic situation.

New York City

DATE · SEPTEMBER 20–21, 1776

DAMAGE · 500 BUILDINGS

CAUSE · ARSON SUSPECTED

The American War of Independence had begun in 1775 with the "shot heard round the world" at Lexington, Massachusetts. In rapid succession the Americans captured Fort Ticonderoga, N.Y., the Second Continental Congress was held in Philadelphia and George Washington was made Commander-in-Chief of American forces. On August 22, 1776 British commander Sir William Howe landed on Long Island and quickly moved to the Kip's Bay section of New York City on the East River. His soldiers occupied the city while Washington moved slightly north into Harlem

Heights. On September 16 British and German soldiers were lured north and surprised by a trap. They fled south through the fields near present day Columbia University, giving the Americans small a victory.

The fire began on September 20 and was rumored to have been set by Americans who supported General Washington and, fearing that the British would take New York, wanted to leave them no shelter. Redcoats attacked suspected arsonists, further antagonizing the

---

ABOVE *The fire was rumored to have been set by Americans who supported General Washington. British soldiers are shown attacking suspected arsonists, further antagonizing the citizens. Engraving by Franz Xaver Habermann.*

FIRE-FIGHTING TECHNOLOGY AT THE TIME

There had been organized fire-fighting in New York since 1648, when the first fire ordinance was adopted by the Dutch in New Amsterdam. These laws regulated building materials and mandated frequent chimney cleanings. Dirty chimneys were a common cause of fires in colonial cities. Fines provided funds for bucket brigades. Fire watches were also established. In 1731 the first fire engines were brought to New York from London and designated Engine Company 1 and Engine Company 2. The volunteer Fire Department was established in December 1737. They had hooks for pulling down buildings, ladders and primitive pumps for water.

citizens. Meanwhile, because of the fighting, the fire spread unabated and with great rapidity through the mostly wooden buildings.

Frantic people threw their belongings out the windows and ran through the streets, heading north towards the safety of the country. Because of the war, many residents wanted to leave no shelter for the British, and thus little of the fire-fighting equipment was put into use. The fire was eventually extinguished by the British soldiers who had taken the city.

ABOVE RIGHT *The cover of a colonial fire insurance policy shows firefighters using a hand-pumped, cart-mounted extinguisher to fight a second-floor fire. Note the line of men on the right filling the pumper with buckets of water.*

RIGHT *New York's first fire engine was the side-stroke, goose neck, tub engine manufactured in England and imported in 1731. It had a water capacity of 170 gallons. The engine was not fitted with apparatus for suction so it had to be filled by buckets. The engine and the company that ran it were dubbed the "Hayseeds" and they were pressed into action by General Washington.*

THE DEED OF SETTLEMENT OF THE Mutual Assurance Company, FOR INSURING HOUSES FROM LOSS BY FIRE IN NEW·YORK.

HAYSEED HUDSON ENGINE 1 NEW YORK

Norfolk, Virginia

DATE · JANUARY 1, 1776
DAMAGE · $1,500,000
CAUSE · MILITARY ASSAULT

"A vast heap of ruins and devastation."
—*Virginia Gazette*, October 11, 1776

Historical Background

The southern colony of Virginia had supported the Massachusetts Colony against the "Intolerable Acts" and the assembly had defied British Governor Dunmore by establishing a Committee of Public Safety. In September 1775 the Tory Governor Dunmore, who had fled Williamsburg, had a fortification built at Great Bridge just outside Norfolk. American patriots led by Colonel William Woodford united with a force of Carolinians under General Robert Howe and defeated the British force at the Battle of Great Bridge on December 9. Dunmore evacuated Norfolk over the next three days and fled to ships, leaving the town in the hands of the American patriots.

On New Year's Day, 1776, Governor Dunmore ordered his ships to bombard the town after hearing that the patriots were insulting him. At about 3:15 p.m., his canons began pounding the town. He also sent small landing parties ashore to burn warehouses near the docks from which snipers had been shooting. The fires increased steadily, whipped by a strong south wind, burning block after block. By daybreak, showers of sparks swept the town and terrified residents fled in panic in carts, on horse or foot along the one road that led out of town.

The fire burned for two days and destroyed over 850 buildings. The entire waterfront and business area of town was reduced to rubble. The only part of the Borough Church left standing were the walls. American General Howe stayed in the ruined city for five weeks. The Virginia Committee of Safety declared that the remaining houses should be burned to prevent their possible use if British troops landed. On February 6, 1776, Howe's troops lit the fires and marched away.

BELOW *The Old Brass Back was the first successful hand-pumped engine, manufactured in New York in 1743. It got its name from the lavish brass on the box of the engine. It saw service under General Washington during the American Revolution.*

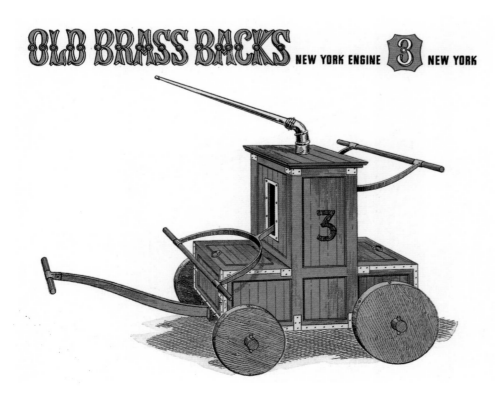

OLD BRASS BACKS NEW YORK ENGINE 3 NEW YORK

The Reaction

The burning of Norfolk convinced most Virginians that there could be no reconciliation with the British and that they must fight for independence. Loyalists found it increasingly dangerous to remain and many fled to Britain. The site was occupied by both British and American troops during the remainder of the war. The rebuilding of Norfolk was slow. There were only 12 houses in 1783, at the end of the war. The census of 1790 showed 2,959 residents, about half the size of the colonial population. As late as 1796 it was still only partially rebuilt with many derelict, half-burnt houses.

Bombay, India

DATE · FEBRUARY 17, 1803
DAMAGE · 482 BUILDINGS
CAUSE · STABLE FIRE

The Great Fire originated in a stable in the Bazaar, probably from an overturned candle, and, fanned by a strong wind, spread with astonishing rapidity.

The *Bombay Courier* of February 19 reported that the fire "was stopped by the prudent precaution of using artillery to beat down the contiguous buildings" thus creating firebreaks. The entire military staff was used to extinguish the fire including the Governor, Admiral Ranier and General Nichols, who directed the fire engines, which were pulled by hand. Two Captains, Elliott and Lane, were injured when a collapsing wall fell on them. Altogether, 471 houses, 6 places of worship and 5 barracks were destroyed.

ABOVE *"Bombay Fort from the Harbour 1665–1700." Nearly a third of the town within the walls had been consumed.*

Nearly the entire native merchant district was destroyed and many buildings were rebuilt on the old foundations. However, in an effort to widen the streets and clear portions of the town following the fire, the government compensated those who were willing to build new houses and businesses outside the walls of the fort.

"A most dreadful and alarming fire…the ravages of which it is impossible to relate."
—*Bombay Courier,* February 19, 1803

The City Before the Fire

Bombay is a port city on the west coast of India and was originally built on seven small islands. Its inhabitants were ruled by the Moslems until the island was ceded to the Portuguese in 1534 by Sultan Bahadur of Gujarat. The Portuguese destroyed Hindu temples and Moslem mosques. The English recognized the value of Bombay as a naval base and landed in Bombay in 1626. The 1650s saw the British East India Company stressing the importance of a port on the west coast of India, which culminated in the marriage treaty of Charles II and the Infanta Donna Catharina of Portugal and placed Bombay in the possession of the English Crown. The building of fortifications to protect it from pirates began in the late 1700s and most residences and businesses were located within the fort walls. It had become so crowded with buildings and slums that, in 1787, a Special Committee was appointed to make recommendations for improvements, which might have taken years if the Great Fire had not indirectly aided their plans.

BELOW *Map of Bombay at the time of the fire.*

The Honorable Jonathan Duncan, in a letter to the Court of Directors, recounted, "So great and violent was the conflagration that at sunset the destruction of every house in the Fort was apprehended. The flames directed their course in a southeasterly direction from the Bazaar down to the King's Barracks. During the whole of the day every effort was used to oppose its progress, but the fierceness of the fire driven rapidly on by the wind baffled all attempts; nor did it visibly abate till nearly a third part of the town within the walls had been consumed."

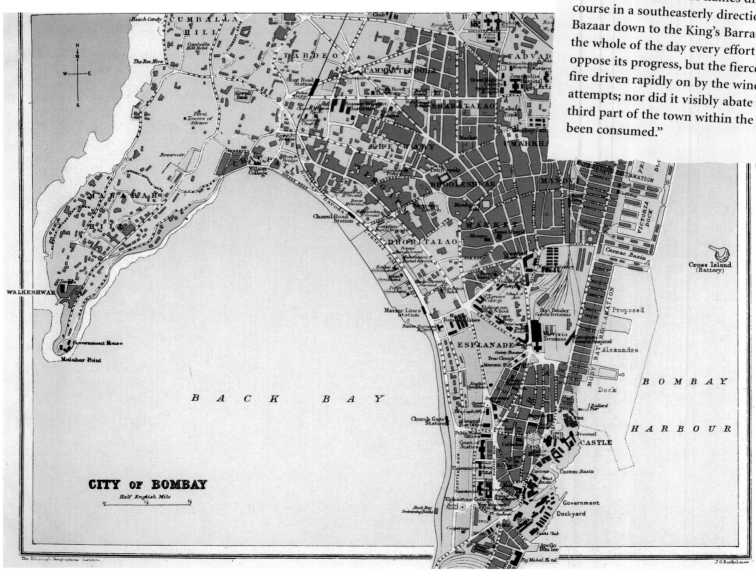

CITY OF BOMBAY

Half English Mile

Austrian Ambassador's Fete, Paris

DATE · JULY, 1810
DAMAGE · 10 DEAD
CAUSE · CANDLE IGNITING A
 CURTAIN

Between 1,200–1,500 guests were present at the reception for Emperor Napoleon I and Archduchess Marie Louise. Dancing took place in a large temporary building made of wood, richly decorated with mirrors, pictures and draperies, and illuminated by a large number of candles. The Emperor Napoleon and Empress Marie Louise were present, along with all the highest nobles of France.

A Scotch reel was being vigorously danced by the Empress, the Princess Borghese, the Princess von Schwartzenberg (sister-in-law to the Ambassador) and about a hundred other women when a candle in one of the sconces near the door fell and ignited a piece of drapery. Colonel de Tropbriant quickly tore down the curtain but this sent sparks flying around the room and ignited other flammables. In less than three minutes the flames had reached the ceiling, which had been smeared with flammable spirit to make it dry quickly.

The Emperor took the Empress by the hand and led her to the garden. There seemed at first to be plenty of time and the rest of the guests started towards the garden entrance without any hurry. However, the heat soon became intolerable, as the flames fed on the hardwood in the room.

Fragments of the ceiling began to fall on the heads and shoulders of the crowd, and, as the pushing increased, people fell and blocked the exit.

The roof collapsed in a fiery crash, sending sparks into the night sky and illuminating the gardens with a ghastly light. Firefighters contained the fire in a few hours by continually pumping water from their hand-pump fire engines, but the devastation was complete.

Only ten people died outright, but hundreds were terribly injured. The affair was regarded as a bad omen for the future of Emperor Napoleon. His humiliating retreat from Moscow occurred just 2 years later, in 1812, followed by his forced abdication and exile to Elba in 1814. His final defeat at Waterloo in 1815, following a brief return, marked the end of his hopes and he was exiled to St. Helena Island in the south Atlantic, where he died on May 5, 1821.

Famous Victims

The Princess de Layen died from her burns within an hour. The Princess von Schwartzenberg was found among the debris, so burned that her body was only recognized by her diamond jewelry. The heat had melted her diadem, and the silver setting had left its mark in a deep groove on her skull.

Historical Background

The marriage of Emperor Napoleon I and Archduchess Marie Louise of Austria was a cause for celebration by both countries. Prince von Schwartzenberg, the Austrian Ambassador to the court of Napoleon, gave a reception at his mansion on the Rue de Montblanc (now Rue de la Chausée d'Antin). The house was surrounded by beautiful gardens in which pavilions had been erected, and where scenes were acted by dancers from the opera, dealing with the early life of her Royal Imperial Highness.

LEFT *The disastrous party thrown by Prince von Schwartzenberg was in celebration of Napoleon's marriage to his second wife, Marie Louise, in 1810. Neither the Emperor nor Empress were injured.*

Richmond Theater, Virginia

DATE · DECEMBER 26, 1811
DAMAGE · 160 DEAD
CAUSE · CANDLE CHANDELIER IGNITING STAGE SCENERY

The first disastrous theater fire in the United States occurred while the Governor of Virginia, George W. Smith and ex-United States Senator Abraham Bedford Venable were enjoying the last act of *Agnes and Raymond*. A careless stagehand raising a chandelier with candles set fire to the scenery. The flames spread rapidly across the flimsy wood and canvass of the set and the estimated 600 people in attendance, panic-stricken, pushed their way towards the single door. The editor of the *American Standard* was at the theater and reported that people were being pushed out of the upper windows by those behind them, who "were seen catching on fire and writhing in the greatest agonies of pain and distress…. One lady jumped out when all her clothes were on fire and she was stripped naked."

LEFT *People were pushed out of the upper windows by those behind them, who "were seen catching on fire and writhing in the greatest agonies of pain and distress." There was only one exit door from the theater.*

The audience members on the main floor mostly escaped, but those in the boxes were trapped by hundreds of people struggling down the one narrow staircase. Dense black smoke suffocated many while the crowd trampled others to death. The fire totally engulfed the building, and in ten minutes the roof and walls collapsed. The Governor and Senator were among those killed.

An ordinance was passed December 28 to allow for the burial of the dead within the ruins of the theater. The citizens purchased the ground on which the theater stood for the purpose of erecting a monument to the victims. The Virginia Legislature passed laws requiring theaters to have multiple exits as well as buckets of water at each aisle.

"Weep my fellow-citizens for we have seen a night of woe, which scarce any eye had seen, or ear hath heard, and no tongue can adequately tell."

—Nathaniel Coverly

The Investigation

The report of the Committee of Investigation found fault with the fact that there was just one means of escape through one door. "The crowd had no other resource than to press through a narrow angular staircase...the loss of so many valuable lives is wholly attributable to the construction of the theater."

Washington, D.C.

DATE · AUGUST 24, 1812
DAMAGE · $2,000,000
CAUSE · MILITARY ASSAULT

Washington, D.C. was captured by British forces under Major General Ross and Admiral Sir George Cockburn on August 24, 1814. The populace had fled in panic and chaos before the invasion and at 3 p.m. Dolley Madison, the young socialite wife of James Madison, had left the President's House with the portrait of President Washington by Gilbert Stuart.

Admiral Cockburn demanded that the Capitol and President's House be burned, and at 9:45 p.m. Ross's men charged up the steps of the Capitol and shot off the locks of the doors. Cockburn sat in the chair of the

"The British have burned and pillaged but posterity is assured of enjoying a determined nation's restoration of these shrines."

—National Intelligencer

Speaker of the House and asked his men, "Shall this harbor of Yankee democracy be burned?" and the resounding answer was "yes." Ross had his men start the fire in the wooden bridge between the wings of the House and Senate. The troops then used books and papers from the Library of Congress, as well as furni-

ABOVE *Dolley Madison, wife of President James Madison, rescues a portrait of George Washington by Gilbert Stuart from the President's House.*

OPPOSITE *British Admiral Sir George Cockburn ordered the Capitol and the President's House burned after taking Washington on August 24, 1812. Troops set fire to the wooden bridge between the House and the Senate and then ignited the President's House. When the President's House was rebuilt, it was painted white and given its current name.*

ture, to start the fires in the House and Senate chambers. Fifteen minutes later tongues of flame could be seen shooting out the windows and moving up to the wooden-shingled roof. The wooden floors and ceiling timbers fed the flames, which became so hot that the marble columns cracked. Within an hour, the Capitol was completely gutted.

The British moved on toward the President's House. There they found an entire banquet on the table, which Dolley had ordered for her husband before fleeing. There was roast beef, lamb, squab, Virginia ham, potatoes and vegetables. There were tables with cakes and tarts and bottles of wine, brandy and port. Cockburn, Ross and their men ate and drank and then took whatever "souvenirs" they wanted. At midnight the British troops set the President's House on fire. They went from room to room with torches igniting draperies and piles of furniture. Several kegs of powder were set off in

the cellar as the men were leaving. Within 15 minutes the entire building was enveloped in flames.

Before retiring for the night, Ross had his troops enter the Treasury Building and take whatever money they could, which was then also set on fire. At 1 a.m. on the morning of the 25th, Ross and his men returned to their camp on Capitol Hill.

On the 25th Ross's men set fire to the State, War and Navy Buildings. They also rekindled the fire at the Treasury Building, which had only partly burned the night before. The Post Office and the Patents buildings were burned, destroying hundreds of models of inventions. The offices of the *National Intelligencer* were burned. They also burned the Arsenal.

About 4 p.m. a tornado struck Washington tearing roofs off houses and toppling huge trees. The British decided to withdraw, and at 9 p.m. they left Capitol Hill after occupying Washington for one day.

James Madison arrived back in Washington on Saturday the 26th and Dolley arrived back on Sunday the 27th. The British next tried attacking Baltimore, which was defended by Fort McHenry, but failed. Francis Scott Key, who was present at the battle, wrote the U.S. National Anthem, *The Star-Spangled Banner*, in response to the American victory.

The Treaty of Ghent, signed December 24, 1814, ended the War of 1812. There was some debate about moving the capital, but through the efforts of citizens' committees and public subscriptions, the capital was rebuilt. The President's House was then painted white and became known as the White House.

Historical Background

The War of 1812 had its beginnings in the Napoleonic wars in Europe. Britain did not want the United States trading with France and began capturing American ships, seizing goods and forcing the sailors to work on British ships. Statesman Henry Clay and Congressman John C. Calhoun persuaded Congress that the British would be engaged in war against Napoleon, and on June 18, 1812 Congress passed a declaration of war against Great Britain, even though the U.S. was totally unprepared for war. At the time, the army of 6,000 was poorly drilled and clothed and the navy consisted of only a few ships.

What was Destroyed

The cornerstone of the Executive Mansion, or President's House, was laid on October 13, 1792. Designed by James Hoban, it was still unfinished when President John Adams and his wife Abigail moved into it in November 1800. Architect Benjamin Henry Latrobe remodeled many of the rooms in a neo-classical style under Jefferson and Madison.

Moscow, Russia

DATE · SEPTEMBER 14, 1812
DAMAGE · 30,800 HOUSES
         LOST; $30,000,000
CAUSE · ARSON

Historical Background

Napoleon Bonaparte was considered by many to be the greatest military man who ever lived. He had crowned himself Emperor of the French in 1804 and then began a war with England, Russia, Sweden and Prussia. His determination to take Moscow led him to the Campaign of 1812. Crossing the Niemen River, he entered Russia on June 24 and took Smolensk with hardly a bullet fired.

ABOVE *The retreating Russians, not wanting to give Moscow to Napoleon, opted to burn it down. The fire was so intense that molten iron and copper ran in the streets, and the ground was so hot it burned through boots.*

A last stand was made against Napoleon by the Russians at Borodino with tremendous losses on both sides. Advancing on Moscow, Napoleon envisioned the Russians asking for peace negotiations, and a grand military entrance into the city with crowds cheering. He did not reckon on the tenacity of the one-eyed General Kutusov, who declared on September 13, "Moscow is not the whole of Russia. Better to lose it than the Army and Russia." He gave the order for the army to retreat from the city. Meanwhile, Count Rostopchin, Governor of Moscow, insinuated to vari-

ous people, "He would burn the capital rather than deliver it to the enemy." On the night of September 13, Rostopchin ordered all fire engines out of the city to follow the army and he left the city on September 14 with the remainder of the fleeing populace.

Napoleon arrived on the Mount of Salutation just outside Moscow on the afternoon of the 14th. He gazed at the city below him and exclaimed with emotion, "So there it is, this famous town!" He had proclaimed months before that "peace lay at Moscow," but was soon to see that this was not true. Moscow was empty and his lonely entry into town was not the type of entry he had planned. One soldier wrote, "I feel chilled with fear not seeing a living soul."

That night, arsonists dressed as peasants, criminals and monks lit fires in all nine districts of the town. Napoleon went from one window of the Kremlin to another watching the flames and the columns of smoke rising from the city.

> "The burning of Moscow has illuminated my soul."
>
> —Tsar Alexander I

The Retreat

During the night of September 17, heavy rain began to fall and Napoleon returned to Moscow on the morning of September 18. Attempts were made at peace, but on October 19 he began his retreat from Moscow. The Russian winter arrived and a lack of food and the cold weather killed many already weak soldiers. Napoleon left the army and returned to Paris on December 18, 1812. Out of his army of 550,000 only 20,000 survived the wintry retreat.

The wind had become a gale and by morning fire was invading the Kremlin. Napoleon finally agreed to leave in the afternoon. The fire was so intense that molten iron and copper ran in the streets and the ground was so hot it burned through boots. The Russian peasants who had not fled were looting along with French soldiers, but many were caught in the flames, which "formed a sky of fire and walls of fire," according to Philippe de Ségur, a French officer at the scene. After a harrowing escape through the flames, which took nearly four hours, Napoleon settled in the Petrovskoye Palace outside Moscow. Montesquieu

ABOVE *When his Russian expedition proves disastrous, Napoleon watches Moscow burn from the windows of the Kremlin.*

wrote about that night, "It was pitch black and the sight was terrifying and grandiose, almost sublime, so closely did it resemble the Apocalypse…. often the flames were drawn aside like curtains and displayed palaces… which appeared to us thus in a fairylike splendor to bid the world a last good-bye."

Houses of Parliament, London

DATE · OCTOBER 16, 1834
DAMAGE · DESTROYED
CAUSE · STOVE OVERHEATING

On the night of October 16, 1834, the old Houses of Lords and Commons and the adjacent buildings were destroyed by a fire, which lit up the London sky and drew crowds of onlookers.

The fire began shortly before 7 p.m., when a stove below the House of Commons overheated while burning ancient wooden tallies, used as financial records by the Exchequer. Firefighters arrived, but the fire spread rapidly, fanned by strong winds. *The Times* of London reported that by 8:30 p.m., "As soon as you shot through [Westminster Bridge] the whole melancholy spectacle stood before you…from the Parliament offices down to

LEFT *Flames were shooting fast and furious through every window of the Houses of Parliament. The fire was so hot and frenzied that efforts to fight the flames directly (as seen here) were useless. However, neighboring buildings, such as Westminster Hall, were saved by dousing them with water before the flames could reach them.*

the end of the Speaker's house, the flames were shooting fast and furious through every window…the roof of the House of Commons had already fallen in." The walls of the House of Lords soon followed, despite the exertions of the firefighters.

The Grenadier Guards, the Coldstream Guards, the Royal Horse Guards and other regiments of the military were ordered to assist the firefighters. By 10 p.m. the fire raged unabated and immense volumes of smoke, sparks and flame poured out of the various parts of the building and were carried on the strong winds in the direction of Westminster Bridge. Firefighters could only direct the water onto neighboring buildings such as Westminster Hall, in hopes of saving them.

At 12:30 a.m. the painted chamber and the library of the Lords were completely reduced to ashes. Westminster Hall caught fire but through the efforts of the firefighters using three hand-pulled pumpers, it was extinguished and little damage occurred.

By 5 a.m. the fire was contained, although firefighters continued to pour water on small flare-ups throughout the morning. Thousands of spectators began to arrive to view the ruins although they were kept behind barriers erected and guarded by hundreds of police and military personnel to protect the ruins from looting. Gawkers included all classes of society, from King William IV to a common dustman. One man exclaimed, "I never thought the two Houses would go so near to set the Thames on fire."

The Parliament Before the Fire

The House of Lords was one of the chambers of the Old Palace of Westminster. The Houses of Commons occupied the former St. Stephen's Chapel, which was supposed to have been originally erected by King Stephen as a chapel in the Old Palace. It was rebuilt by King Edward I in 1292, and again in 1330 by Edward III. The complex also included Westminster Hall and Westminster Abbey.

"There was an immense pillar of bright, clear fire and cloud of white, yet dazzling smoke careening above it through which, as it was parted by the wind, you could occasionally see…lanterns…and the clear stone of Westminster Bridge which stood out well and boldly in the clear moonlight."

—*The Times*

The Times reported, "The spectacle was one of surpassing though terrific splendor." The Thames, the Bridge and the opposite shore were soon crowded with thousands of spectators. "The conflagration, viewed from the river, was peculiarly grand and impressive."

New York City

DATE · DECEMBER 16–17, 1835
DAMAGE · 2 DEAD; 674
BUILDINGS;
$20,000,000
CAUSE · GAS LEAK

The night of December 16, at 17 degrees below zero, was believed to be the coldest night in thirty-six years. Two feet of snow had fallen and in addition, an icy gale was blowing from the north.

The largest fire in the United States to date was discovered by Peter Holmes, a watchman, shortly after 9 p.m. at Comstock and Andrews Dry Goods store at 25 Merchant Street. It was later discovered to have been caused by a leaking gas pipe, ignited by a coal fire.

The alarm was sounded immediately, but in less than a half hour both sides of the street were ablaze. Firefighters were unable to get water to the flames because the extreme cold weather had frozen the water in the hydrant pipes and wells, so the fire spread rapidly, blown by the wind. According to the *Evening Post* of December 17, by 11:30 p.m. "the flames are now raging in every direction…three or four of the buildings in Exchange Place are on fire and the whole block on

---

BELOW *Map showing the burned district, which included a 17–block area surrounding Wall Street.*

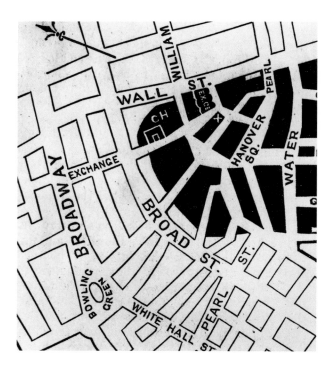

The Fire Department at the Time

The fire department consisted of 56 engines, 6 hook and ladder carts, 5 hose carts, 15,000 ft. of hose and 1,500 volunteers. The fire engines were pulled to the scene by the firefighters and hand-pumped to draw water. Steam-driven fire engines drawn by horses were not introduced until 25 years later.

William Street, as well as the Exchange." By 1 a.m. "the hydrants are exhausted, the hoses of many of the engines are frozen and useless."

By 3 a.m. the Exchange was in ruins and the Post Office was destroyed. Mayor Cornelius W. Lawrence decided after consultations to blow up buildings on Exchange Place to stop the progress of the flames. Charles King, editor of the *American* and later President of Columbia College, rode to Governor's Island in an open boat, despite the gale, to get the powder. At dawn on the 17th, the explosives began to be deployed.

Merchants rushed downtown to try to save their goods, and horses and carts were commandeered to haul merchandise. Much of it was placed in the Merchant's Exchange and the Garden Street Dutch Church, only to be burned as the fire grew. Liquor stores tried to move their supplies, but hordes of criminals (and thieves) had streamed into the area pilfering goods in the mayhem. Drunken men and women were soon reveling in the destruction, looting houses and stores. More than four hundred people were arrested for theft.

One merchant, dubbed "The Oyster King," saved his Broad Street store by applying pails of vinegar to the roof and walls, in place of water. A lead merchant later found his entire stock melted. An attempt to save a marble statue of Alexander Hamilton from the Merchant's Exchange was abandoned when the flames approached. It was reported that, as the Garden Street Church burned, someone began playing a dirge on its organ, which ended abruptly when the roof collapsed.

The fire was finally controlled by the afternoon of the 17th by blowing up entire blocks and the stunned city flocked to see the ruins. The gloom was reflected

ABOVE *A view of the ruins after the Great Fire in New York, December 16 and 17, 1835. It was the worst fire in the world since the London fire of 1666. Losses were estimated at more than $20 million.*

by James Gordon Bennett, writing in the *Herald*, "Good God! In one night we have lost the whole amount for which the nation is ready to go to war with France!"

Within a few days, optimism began to grow and plans were made for erecting new buildings and widening the streets. Editorials in newspapers advocated various reforms, such as using a different mortar mixture, piping water from the rivers, use of iron for safes and fire patrols.

One year after the fire Philip Hone, famed New York diarist, wrote, "The whole is rebuilt with more splendor than before…and all this with no actual effectual relief from the general or state governments."

The Damage

The losses were enormous. There were 674 buildings destroyed and thousands of clerks, porters, cart men and laborers were out of work. The total amount of damage was believed to be approximately $20,000,000, which in turn bankrupted 14 insurance companies. Fortunately, because the fire was confined to the business district, there were only two deaths.

Lexington Steamship

DATE · JANUARY 13, 1840
DAMAGE · 140 DEAD
CAUSE · CARGO BAY FIRE

"This is by far the most distressing steamboat disaster which has ever occurred in Long Island Sound, or indeed in this portion of the Union. The sufferings of that awful night can never be described nor conceived."

—*Bridgeport Farmer Extra*

The steamship *Lexington* left New York at 3 p.m. for Stonington, Connecticut. About 7:30 p.m. off Eaton's Neck, Long Island, fire was discovered in the ship's cargo of cotton bales. They had been piled against a pipe that led from the boiler-fire room through the freight area to funnel exhaust above. The heat from the pipe had ignited the bales. An alarm was immediately given, but all efforts to subdue the flames using the fire hoses onboard proved useless—the cotton was burning too rapidly, producing intense heat. The flames quickly grew into a raging inferno, so the pilot headed the boat toward the Long Island shore. However, about fifteen minutes later, the rudder ropes burned in two, and the ship lost its steering. Meanwhile the engine kept going, under a heavy head of steam.

When the ship was within two miles of the shore, the engine stopped suddenly, due to a mechanical breakdown that was probably caused by the fire, though investigators were never able to ascertain this for certain. Three lifeboats were launched, with about twenty people in each, but were swamped because they were overloaded and were launched bow-first into the water, causing them to tip over and toss the terrified passengers into the frigid water. A fourth boat was crushed by the paddle wheel.

The fire was raging in the center of the ship, cutting off communication between the fore and aft. Panicked, screaming passengers crowded together on the bow and stern until forced to hurl themselves overboard to escape the flames. The crew threw unburned bales of cotton overboard in the hope that they would provide something on which passengers could float.

The Problem with the Steam Engine

Steamships of this era used extremely hot coal fires to create the steam which powered the boats. The fires, contained in boilers, reached over 212 degrees to boil the water, which produced the steam that was sent through pipes to the cylinders. The steam expanded under pressure, and was admitted into the cylinder by a valve mechanism. In turn, the steam in the cylinders pushed the piston, which was connected to a crank on the flywheel, which propelled the boat. Fire was a common problem with steamboats of this period. The high heat frequently burned through surrounding wood, or at other times burned away the water in the boiler, causing an explosion.

Captain Hilliard remained onboard until the flames reached the wheelhouse from the promenade deck. He and a passenger lowered the last bale into the water at about 8 p.m. and sat astride it. They looked back at the flaming wreck, which burned until it sank at about 3 a.m. They drifted among floating bodies throughout the night. About 4 a.m. the passenger fell off the bale and sank wordlessly into the black water. Captain Hilliard floated on the bale for 15 hours until rescued by the sloop *Merchant*.

The Survivors and Victims

Captain Hilliard was one of the only three survivors. All the other crew members and passengers either died in the fire or drowned in the icy waters of Long Island Sound. Wreckage from the *Lexington* washed up on the North Shore of Long Island for weeks afterwards.

Hamburg, Germany

DATE · MAY 5–8, 1842
DAMAGE · 50 DEAD, £7,000,000
CAUSE · UNKNOWN

ABOVE *Residents fleeing with furniture and household goods abandoned them in the streets when the flames closed in, blocking efforts to fight the fire.*

The City Before the Fire

Located on the Elbe River in northern Germany, Hamburg was a leading commercial port of the Hanseatic League. Briefly under the control of Napoleon, it flourished as a "Free City" until its incorporation into the German Empire in 1871. It had a population of about 130,000 people in 1842.

The fire was discovered at about 12:30 a.m. on Thursday, in a tobacconist's shop on Deutschstrasse. Most of Hamburg's stores, warehouses, office buildings and homes were old and primarily built of wood. Because it was the Feast of the Ascension holiday, firefighters were slow to respond, and twenty buildings had been destroyed before they arrived.

A strong wind from the southwest fanned the flames, and inhabitants began to flee with furniture and household goods, which were abandoned in the streets when the flames became too strong. The firefighters'

efforts were blocked by the panicked crowd. Thus, the fire spread unchecked.

By Friday the fire had spread northward beyond the Hopfenmarkt and upwards of 300 buildings were destroyed. From 10:30 a.m. to 8 p.m. that day over 400 more houses were destroyed. The destruction continued unabated through Saturday. Firefighters from Mecklenburgh, Lubeck and other towns arrived to assist, but were unable to contain the fire with their water hoses. Late Saturday, the Senate finally gave the order for troops to blow up buildings. *The Times* of London reported that three English naval engineers, Lindley, Giles, and Thompson, volunteered to lay out the explosive powder, but were mistakenly arrested as they were setting it, on suspicion of being arsonists.

> "Germany will certainly not fail to lament the destruction of this great and flourishing town, the acknowledged seat of science and knowledge, and will hasten to extend its brotherly assistance to the unfortunate citizens."
>
> —*Hamburger Neue Zeitung*

They were quickly released and were able to blow up the buildings, which served as a firebreak, stopping the spread of the fire and allowing the firefighters to use their hoses to extinguish what was left burning.

No church services took place on Sunday because the buildings were housing the estimated 30,000 homeless. Over 2,000 houses had been destroyed.

Hundreds of laborers pulled down burnt walls and cleared rubbish. Plans were made to broaden the streets and fill in some of the canals. Monetary subscriptions for the relief of the victims were begun with contributions from the Kings of Prussia, Denmark and Hanover, as well as Queen Victoria and the Duke of Mecklenburgh.

The Damage

At least three churches, including the St. Nicholas, burned, and their spires, some 400 feet tall, became firebrands shooting flames and sparks over the city. The City Hall burned, as did numerous banks, although the New Exchange was saved.

The Times of London provided figures on the quantities of goods destroyed by the fire:

Coffee	2,000,000 lbs.
Sugar	5,000,000 lbs.
Cotton	1,550 bales
Carolina rice	300 tons
Java rice	500 sacks
Palm oil	1,000,000 lbs.
Wheat	2,000 tons
Wine	8,000 barrels
Brandy	400 casks
Tobacco	3,000,000 lbs.

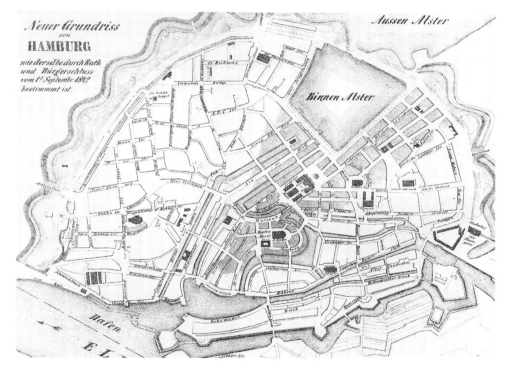

TOP LEFT *The St. Nicholas Church steeple went up in flames and became a firebrand shooting flames and sparks over the city.*

TOP RIGHT *Firefighters were able to pump water directly from the Elbe River, through the use of ship-mounted pumps.*

ABOVE *Map of Hamburg with the burned area in grey.*

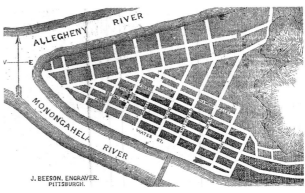

ABOVE *In this map of the burned district, the black area, from Ferry Street past Ross Street, marks the burned area, which was about one-third of the city.*

LEFT *"The Great Conflagration at Pittsburgh, PA, April 10, 1845," an engraving by James S. Baillie. The Monongahela Bridge (center), connecting downtown with the South Side, burned in just 8 minutes.*

Pittsburgh, Pennsylvania

DATE · APRIL 10, 1845
DAMAGE · 2 DEAD; $5,000,000
CAUSE · STRAY SPARK FROM AN OPEN FIRE

"It was a sad day for Pittsburgh. All its best business houses in ruins, hundreds of dwellings wiped out, presented a sad and sickening sight."

—Samuel Young, newspaper editor

The City Before the Fire

Pittsburgh is situated on a triangular shape of land bounded by the Monongahela and Allegheny Rivers, which converge at "The Point" to form the Ohio River. Fort Duquesne, built by the French in the 18th century, became Fort Pitt after it fell to the British. The town of Pittsburgh was incorporated in 1816.

The Spring of 1845 had been unusually warm and dry. It had not rained for six weeks, and drought conditions prevailed. Fierce winds had been gusting from the West. At 12:05 p.m. on the morning of April 10 a washerwoman decided to do her laundry outside, and lit a fire in the backyard of the home of Colonel William Diehl, at Second Avenue and Ferry Street. A gust of wind carried a spark to the barn, which quickly ignited. Next door, Bruce's Icehouse was rapidly ignited by sparks from the barn and then the whirling wind carried more sparks, igniting the cotton factory of Colonel James Woods, across the street. The Third Avenue Presbyterian Church stood at the corner of Ferry and Third, and the bells were rung furiously to signal fire.

There were six competing volunteer fire companies in the city proper: the Niagara, the Neptune, the Duquesne, the Vigilant, the Allegheny and the Eagle, with four more in Allegheny City—across the Allegheny in what is now called the North Side—the Uncle Sam, Washington, William Penn and Union Hose. However, there were just two water mains in town, along Liberty Avenue and Third Street. The reservoir on Quarry Hill was so low that the fire engines sucked up only mud. The engines had to be hand-pumped, and there was not enough hose to reach from the rivers into the downtown district. Volunteers from all parts of the city responded to the call, but the strong wind carried the flames from building to building faster than the volunteers could respond.

The Third Avenue Presbyterian Church roof caught fire, but it was saved and acted as a firebreak, preventing

The Relief

News of the disaster reached Pennsylvania's capital on the 12th, and the Legislature appropriated $50,000 for the relief of the sufferers, which was distributed under the direction of the Mayor and Council. Donations amounting to over $200,000 came from Europe, as well as notable Americans, such as President James Buchanan, President John Quincy Adams and Edwin M. Stanton.

the fire from spreading in a northerly direction. The fire turned towards the east, burning the Bank of Pittsburgh on Fourth Avenue and melting its zinc roof. The fire then sped up Third and Second Avenues toward Market Street. From there it moved along Smithfield Street towards the Monongahela wharves, piled with goods, and onto the warehouses.

By 7 p.m. the fire had burned itself out. Historian Leland D. Baldwin tells us "Twenty squares in the most valuable part of the city, comprising about fifty-six acres and with nearly 1,000 buildings, had been ravaged and 12,000 people had been rendered homeless. About one-third of the area of the city was in ashes, and approximately two-thirds of its wealth had been destroyed." Only two people were killed. Between 975 and 1,115 buildings were left in ruins.

Within a couple of months, new, larger commercial buildings had risen and the city was booming with trade for the West. As a direct consequence of the fire, the Firefighters' Association of Pittsburgh and Allegheny was formed in August to "promote good order, efficiency and harmony" between the companies of both cities.

From 2–4 p.m. the fire raged at its height. The *Pittsburgh Gazette* wrote on April 11, "Huge waves of flame rolled on and on, burying in their swelling tide the homes and hopes, the prosperity and fortunes of the orphan, the widow, the wife, the husband, the mechanic, and the princely merchant in one common grave." The buildings of the Western Pennsylvania University were rapidly destroyed, as was the Monongahela House Hotel. According to Charles Arensberg's article in the "Fire Centennial" issue of *The Western Pennsylvania Historical Magazine*, March-June 1945, Thomas Mellon, later Secretary of the Treasury, "walked up Third Avenue to Smithfield, among people running in all directions, wild with excitement, with cinders and burning shingles falling everywhere."

BELOW *A view of Pittsburgh after the fire. About fifty-six acres, with nearly 1,000 buildings, had been destroyed, and 12,000 people had been left homeless.*

"All the masts were gone, all the decks aft and fore were in flames and fire was bursting in holes through the sides."

—First mate J. Bragdon

The *Times* reported, "The women and children received every attention; the children were carefully looked after by the Princess and Duchess." Another steamer, *The Prince of Wales*, saved 17, while 13 more were saved by another ship. The total rescued was 218 passengers.

Ocean Monarch

DATE · AUGUST 24, 1848
DAMAGE · 178 DEAD
CAUSE · COOKING FIRE

The *Ocean Monarch* sailed with a light breeze on the Mersey River at daylight. There were 396 persons onboard, which included 322 steerage passengers, 32 cabin passengers and 42 crew. At about noon, while sailing in the Irish Sea, near the Great Ormshead on the north coast of Wales, the steward reported to Captain Murdock that a fire had begun in steerage. Apparently, a steerage passenger had started a cooking fire in a wooden ventilator on the third deck.

ABOVE *This dramatic image of the burning* Ocean Monarch *was drawn by Morel Fatio, from the sketchbook of the Prince de Joinville. The Prince was a passenger, with his wife, on the Brazilian steamer* Allonzo, *which rescued 156 passengers. Sea rescues like this are very dangerous, as large boats try to maneuver close enough to pick up overboard passengers without catching fire.*

In five minutes the aft portion of the ship burst into flames, which were fanned by the breeze across the moving ship. Attempts to put out the fire with water were futile, and the Captain faced the ship into the wind in

order to lessen the flames. According to an account by Captain Murdock in *The Times*, "The fire produced the utmost confusion among the passengers…yells and screams of the most horrifying description were given…all control over them was lost…my voice could not be heard." Two lifeboats were lowered, before flames devoured the others. The ship's masts and spars were soon wrapped in flames and crashed to the deck, showering many helpless victims with burning debris. The fire was bursting with immense fury from the stern and the center of the ship. Many passengers rushed to the bow to avoid the heat of the flames. In despair and fear, women jumped overboard, with their offspring in their arms, and men followed their wives, in a frenzied hope of saving them. Captain Murdock abandoned ship just before the flames would have trapped him. He clung to a board until picked up by the *Queen of the Ocean*, the first boat to the rescue.

Testifying at the inquiry afterwards, no fault was found with the conduct of the crew—in fact, they were praised for their tireless efforts to rescue the passengers. Contributions for the survivors poured into charities, including a large sum from the Prince de Joinville and the crew of the *Allonzo*.

A Hero

By 3 p.m. the ship was burnt down to the water, and only a dozen helpless women and children remained on the burning wreck. An English seaman, Frederick Jerome proved himself a hero and, "stripping himself naked, made his way through the sea and with a line in his hand succeeded in lowering the last helpless victims safely into the boats, being himself the last man to leave the wreck."

Sacramento, California

DATE · NOVEMBER 2, 1852
DAMAGE · 7 DEAD; $10,000,000
CAUSE · UNKNOWN

The City Before the Fire

Sacramento lies on part of a Mexican land grant originally belonging to John A. Sutter, who, in 1839, began a settlement called New Helvetia. The discovery of gold in 1848, at Sutter's Mill, started the California Gold Rush, and the thousands of men known as "The 49er's," seeking their fortunes in the California gold fields, settled here. Sacramento grew rapidly into a town of 10,000 people, and was incorporated as a city in 1850. That same year a fire destroyed three hotels, a store and several wood and canvas buildings. It was at this time that a volunteer Fire Department was organized.

The Great Fire of 1852 was discovered at 11:10 p.m., when flames and dense smoke were seen coming from the millinery shop of Madame Lanas, on the north side of J Street. An alarm was sounded and the volunteer firefighters responded quickly.

Despite their efforts, the fire spread rapidly across the street to the Crescent City Hotel, blown by an ill-timed gale from the north. In less than ten minutes the fire had burned through K Street and Fourth Avenue. It spread east, destroying the Phoenix Hotel, the Roman Catholic Church, the Baptist Church and the Methodist Church. It moved south, and burned through other hotels, office buildings and homes. Just after the fire, the *Daily Union* recounts, "The wind…increased to a hurricane, carried burning boards several feet in length through the air, and lighted the roofs of buildings often one or more blocks in advance of the main conflagration."

After burning for seven hours, the fire was finally extinguished. The morning of the 3rd revealed that the entire business district was demolished and hundreds were left homeless.

Fire Behavior

Wind can foil even firefighters' most efficient efforts by carrying sparks to adjacent buildings and starting spot fires as far as 500 yards away. Hurricane-strength wind vortexes caused by wildfires have been known to carry sparks for miles.

Relief from other northern California cities, such as San Francisco, was immediate. Food and building supplies arrived, and within a month most of the houses had been rebuilt using wood, with the business district using brick.

ABOVE *Firefighters try to extinguish the fire in Sacramento.*

The *New York Daily Times* reported on December 13, "Out of twenty-eight hundred buildings, scarcely two hundred remained the morning after the conflagration."

Milwaukee, Wisconsin

DATE · AUGUST 25, 1854
DAMAGE · $500,000
CAUSE · UNKNOWN

ABOVE *The volunteer fire companies struggled to halt the spread of the flames. At the height of the fire one of the city's five hand-pumps had to be pushed into the river to save it.*

The Fire Department at the Time

Milwaukee had approximately 30,000 residents in 1854. The first volunteer fire department was formed in 1837, and in 1839 the first fire engine was acquired, for Engine Company 1. On April 7, 1845 a fire began in a warehouse storing gunpowder, which exploded and subsequently burned an entire block. The single fire engine's inadequacy in controlling this fire showed that more and better fire-fighting equipment was needed. New engines were purchased, and additional volunteer companies were formed.

The most destructive fire in Milwaukee history occurred on August 24, 1854. It originated in the Davis livery stables, in the rear of the U.S. Hotel, at about noon. The cause of the fire remains unknown. The fire soon spread through the entire block of wooden buildings bounded by Huron and Michigan streets, and then crossed East Water Street. It swept through the stores occupied by Bosworth & Sons and Haney & DeBow. The heat was intense and thick, black smoke poured from the fronts of the buildings. Fierce northwest winds drove the fire onward, and it soon engulfed the Wisconsin Marine and Fire Insurance Company Bank, at East Water and Michigan Street. It also destroyed the offices of the *Daily Wisconsin* newspaper.

The volunteer fire companies were powerless to halt the spread of the quickly moving flames, which were fanned by the gusting winds, and at the height of the fire one of the five hand-pumps, which was being used to save the Bosworth Building, was in danger of being engulfed in flames, and was pushed into the river to save it.

The fire had been confined to the business section of town, and rebuilding started immediately. In four hours the fire had caused over $500,000 in damage.

The Call for Help

By 2 p.m., Mayor Kilbourn had to telegraph to Racine for help. Two hundred men responded to the call, but by the time they arrived, two hours later, the fire had burned itself out in a thinly settled part of the city.

BELOW, LEFT & RIGHT *East Water Street in 1844. The strong northwest winds blew the fire across this street and through the stores occupied by Bosworth & Sons and Haney & DeBow.*

BOTTOM *Hand-pulled and pumped fire engines are used to battle a building fire. It took many men to keep up the exhausting work of pumping water through the hoses and maneuver the heavy engines.*

"Why do they drag their engines to fires, when horses would do it so much quicker, besides leaving the energies of the men untaxed, in readiness for their legitimate employment?"

—*Milwaukee Sentinel,*
August 25, 1854

Crystal Palace, New York City

DATE • OCTOBER 5, 1858
DAMAGE • DESTROYED
CAUSE • UNKNOWN

The plans for a New York World's Fair began in 1852, with the organization of an "Association for the Exhibition of the Industry of All Nations." The Association held a competition for design of an iron and glass exhibition hall, to be built on 42nd Street, next to the Croton Water Reservoir, on the present-day site of the New York Public Library and Bryant Park.

Many well-known architects of the time sent proposals, and the winning architects were Georg J. B. Cartensen, who had designed Tivoli Gardens in Copenhagen, with Charles Gildmeister, a German architect based in New York.

The fire that destroyed the Crystal Palace began in the afternoon, during the annual fair of the American Institute of New York. It started in some bales of hay in the lumber room, on the south side of the building.

"The prevailing style of the architecture is Moorish and Byzantine in its decorations."

—Horace Greeley

BELOW *An interior view of the Crystal Palace before the fire. The two-story building had a dome at the center and was built largely of pre-fabricated iron components. It was reputed to be fireproof, due to the use of the iron and glass, although it did have wooden floors.*

OPPOSITE *Burning of the New York Crystal Palace. The heat was so intense that the glass and iron floors and roof soon collapsed. The entire building was completely destroyed in fifteen minutes.*

The Building Before the Fire

In August 1852 construction of the 451 foot long central arcade of the Crystal Palace began. The two-story building was in the shape of a Greek cross, with a dome at the center, and was built largely of prefabricated iron components. It covered 173,000 square feet, and the cast iron weighed 1,200 tons, and the wrought iron 300 tons. It was reputed to be fireproof, due to the use of the iron and glass, although it had wooden floors. On July 14, 1853 President Benjamin Pierce opened the Crystal Palace. The fair had more than four thousand exhibitors from around the world, showing agricultural products, industry, sculpture and paintings. Initially, the fair drew crowds, with paid attendance exceeding one million, but the Association was left in debt when the fair closed on November 1, 1854. The Crystal Palace was leased to various groups during the following years.

The Advent of the Crystal Palace

The great Crystal Palace, built in London by Joseph Paxton, and opened on May 1, 1851 by Queen Victoria and Prince Albert, became an instant success. The breathtaking iron and glass building inspired a number of other Crystal Palaces around the world, including ones in Amsterdam, Dublin, Sydney and the New York Crystal Palace. (*See also* entry on the burning of the London Crystal Palace, 1936).

THE APPEARANCE OF THE CRYSTAL PALACE THE MORNING AFTER THE FIRE.

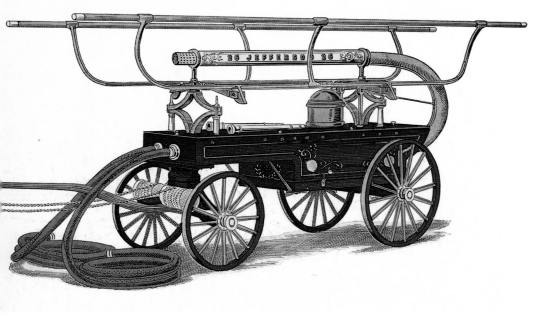

THE BLUE BOYS   JEFFERSON ENGINE **26** NEW YORK

LEFT *The appearance of the Crystal Palace the morning after the fire. It is now the site of the New York Public Library and Bryant Park.*

BELOW, LEFT *Known as the "Piano Box" for the shape of its holding tank, this engine, built in 1853, had suction hoses permanently attached to the back for taking in water. Engines equipped in this manner were known as "Squirrel Tails." The pumping levels folded up (as seen here) so the machine could be maneuvered through narrow passages. Once in position, the levers would be lowered to breast height for pumping.*

When the fire was first discovered, the Palace fire engine was brought out, but the hose had rotted, and was so full of holes that it was useless. At the time, fire inspectors were not as diligent checking rescue equipment in private buildings as they are today. Today, inspectors in most towns are supposed to check fire prevention equipment, such as hoses and extinguishers, once a year. The flames spread rapidly to the exhibits and merchandise, as well as the wooden floors. The building was quickly engulfed in a roaring fire. The estimated 2,000 visitors rushed panic-stricken for the exits.

New York City firefighters arrived a few minutes later, but their efforts were of no avail.

The heat was so intense, from the wood floors and other debris burning, that the glass, iron floors and roof soon collapsed. The entire building was completely destroyed within fifteen minutes.

The Building Material

Cast iron is an alloy of iron containing 2–4 percent carbon, along with varying amounts of silicon, manganese and trace elements. Iron is melted in a blast furnace and cast into molds. It was used extensively as a structural metal in the 19th century because its compressive or load-bearing, strength is good. However, it is brittle and its tensile strength is poor. Exposed cast iron performs poorly during a fire, usually fracturing and collapsing relatively quickly. Today, safer, more fire-resistant steel construction has replaced iron. Structural steel softens and bends, instead of fracturing.

Troy, New York

DATE · MAY 10, 1862
DAMAGE · 5 DEAD, $2,500,00
CAUSE · SPARK FROM A
LOCOMOTIVE

"Disastrous fire in Troy"
—New York Times headline

The "Great Conflagration of 1862" began at noon on May 10, when the shingle roof of the Rensselaer and Saratoga Railroad bridge across the Hudson caught fire from the sparks of a passing locomotive. Firebrands were carried by high winds, and swept southwest over the heart of the business district. The fire quickly burned from Bridge Avenue south to Federal Street. The progress of the flames was so rapid that, within an hour, a belt of fire stretched across the city, from the river to the eastern hills.

Using new steam fire engines, which had been bought the year before, hundreds of firefighters and citizens fought the fire, assisted by volunteer fire companies from other towns. Still, the inferno destroyed 507 buildings, including the Second Presbyterian Church, the North Baptist Church, the Home Mission, Rensselaer Polytechnic Institute, the Troy City Bank, the Troy Orphan Asylum and Union Station. The firehouses of the Hugh Rankin Engine Company and

The City Before the Fire

Troy, New York, was originally part of the patroonship given to Killiaen Van Rensselaer by the Dutch West India Company. The town was laid out in 1786, on the east bank of the Hudson River, at the mouth of the Mohawk River. The first large fire in Troy occurred in 1820. It raged for six hours and destroyed 90 buildings, which amounted to one-ninth of the total buildings in town. On August 25, 1854 another fire spread over seven blocks, and destroyed over 200 buildings resulting in a property loss of $1,000,000.

The Covered Bridge

Covered bridges became popular in the United States in the 18th century. They were timber-truss structures, in which the basic form of support was a triangle that allowed engineers to span rivers without constructing supporting piers. The roofs and siding protected the wooden frame from the weather. The first long covered bridge in the U.S. was built by Timothy Palmer, over the Schuylkill River at Philadelphia in 1806, and was 180 feet. Ithiel Town, from New Haven, Connecticut, patented the "Town lattice" truss, which was sold by the foot and used for many bridges throughout New England. Because the bridges were constructed entirely of wood, they frequently burned, but were usually in remote locations, where they did not threaten a town.

ABOVE *The Green Island Bridge (pictured here shortly before the fire) was set aflame by sparks from a passing steam locomotive. The wind carried blazing shingles, which started the fire in the city.*

The Damage

The fire destroyed 75 acres. Fortunately, only five people, including one child, lost their lives. Hundreds of people were left homeless, but help from neighboring cities provided the necessary shelter and food.

The Fire Found Still Burning a Year Later

On May 9, 1863, a year after the fire, a smoldering fire was discovered, still burning in a covered heap of coal in the cellar of Dusenberry & Anthony's store on River Street.

Trojan Hook and Ladder Company were also destroyed.

By 6 p.m., the efforts of the firefighters checked the spread of the fire at Donohue & Burges's carriage factory, at Seventh and Congress Streets.

The work of rebuilding began immediately. By November, six months after the fire, all the lots had new buildings, except two.

TOP *The ruins of the fire from the western perspective.*

ABOVE *Troy's "Arba Read" No. 1, delivered in March 1860, was the pride and joy of the city. The pump was driven by a one cylinder steam engine which worked alongside hand pumped engines and was very advanced for its time.*

RIGHT *The steamer used by the Hugh Rankin Steamer Co. No. 2 to fight the Great Fire at Troy. The engine house was destroyed in the fire.*

BELOW *Ruins of Troy after the Great Fire of May 10, 1862, showing Union Depot (center) with the Hudson River in the background. Firefighters checked the spread of the fire at Seventh and Congress Streets, after it had destroyed 75 acres.*

THE FIRE DEPARTMENT AT THE TIME

The first volunteer fire department in Troy was organized in June, 1796, with one hand-pumped engine. Hook and ladders came in 1813. Following the 1820 fire, water hydrants replaced well pumps. In 1859 the department consisted of five engineers, 461 members in 12 engine companies, 90 members in 3 hose companies and 83 members in 2 hook and ladder companies.

Atlanta, Georgia

DATE · NOVEMBER 12–15, 1864
DAMAGE · DESTROYED
CAUSE · MILITARY ASSAULT

The City Before the Fire

By the 1850s, Atlanta was one of the South's major trading cities, with railroad and telegraph connections to the nation. New hotels, such as the Atlanta Hotel, The Gate City Hotel and the Trout House provided magnificent rooms to visitors. The city boasted theaters such as the Athenaeum, and a brick City Hall and Court House. Large department stores such as Norcross provided luxury goods. There were 13 churches, 17 commercial buildings and 6 banks.

"Atlanta is ours and fairly won."
—General Sherman in a telegram to President Lincoln on September 2

The bombing of Atlanta had begun earlier in the summer, but on August 9, 1864 General Sherman's big guns had arrived from Chattanooga. They blasted the city with an estimated 5,000 bombshells. Columns of smoke rose upward as the buildings blazed and collapsed.

Five hundred volunteer firefighters, led by chief Thomas Haney, tried to extinguish the flames, but the embers from one blazing building quickly ignited another, until it seemed the whole town was ablaze. The firefighters fought the towering flames, and were successful in extinguishing them using firebreaks and tons of water. Bombing continued day and night through August. General Sherman declared, "Let us … make it a desolation." His intention was to make Atlanta unable to wage war.

The night of September 1, Confederate troops, under orders from General John Bell Hood that "nothing be left in the city to comfort or aid the enemy," began to burn all the supply depots in town. They set fire to the munitions depots, and even the trains and locomotives. Wallace Reed wrote, "It took five hours to blow up the seventy carloads of ammunition. The flames shot up to a tremendous height and the exploding missiles scattered their red-hot fragments right and left. The very earth trembled…the houses rocked like cradles and on every hand was heard the shattering of window glass and the fall of plastering and loose bricks." The few stragglers who had remained in town raced through the streets to evade the fire and the

ABOVE *William Tecumseh Sherman, a Union general, ordered the burning of Atlanta. Houses, railroad depots, factories, foundries and mills were destroyed.*

approaching Union troops. On September 2, Mayor Calhoun surrendered the city.

Following a forced evacuation of Atlanta residents in September, Sherman declared "Leave nothing standing." Atlanta was to be destroyed. The chief of engineers, Colonel Orlando Poe walked through the city marking buildings to be destroyed, followed by Colonel Charles F. Morse planting explosives and kerosene-soaked rags in the buildings.

The fires were lit on Saturday, November 12. Over the course of four days every business, commercial and military building burned. Houses and churches were supposed to be safe, but burning embers rained down on the roofs and spread the flames through the residential district. It was a tremendous sight. "Clouds of heavy smoke rose like a pall over the doomed city," according to one eyewitness. Captain George Pepper reported "Every house of importance was burning…railroad depots, rebel factories, foundries and mills were destroyed. This is the penalty of rebellion."

The night of the 15th the unchecked fire reached its climax.

The Movie

The "Burning of Atlanta" scene from *Gone with the Wind* was filmed using old sets on the MGM backlot. Art Director Lyle Wheeler had the sets from *Little Lord Fauntleroy*, *The Garden of Allah* and *King Kong* burned in December 1938 to make way for new Atlanta sets.

Colonel Underwood described the scene: "A great glare of light from acres of burning buildings…. the intermittent explosions of powder, stored ammo and projectiles, streams of fire that shot up here and there from heaps of cotton bales and oil factories, the crash of falling buildings, and the change, as if by a turn of the kaleidoscope, of strong walls and proud structures into heaps of desolation; all this made a dreadful picture of the havoc of war, and its unrelenting horrors."

ABOVE *Ruins of the Old Kimball House.*

At dawn on the 16th, with the ruined city still smoldering, Sherman assembled 62,000 troops and left Atlanta, with his troops singing the *Battle Hymn of the Republic* with its chorus of "Glory, Glory Hallelujah" for his "March to the Sea."

Richmond, Virginia

DATE · APRIL 3, 1865
DAMAGE · $30,000,000
CAUSE · MILITARY ASSAULT

Just before dawn on Monday the 3rd the remaining residents of Richmond were awakened by terrific explosions. The powder magazine at the arsenal had been blown up when retreating confederate soldiers ignited hundreds of rounds of shells. Wooden ships loaded with explosives, anchored in the river, were detonated, setting the bridges on fire. The warehouses of cotton, tobacco and other goods that would be of value to the Union were set on fire. Most citizens had already left the city, and the remaining ones were not considered important enough to be warned.

Union Lieutenant R. B. Prescott described the scene: "Richmond was literally a sea of flame, out of which the church steeples could be seen protruding here and there, while over all hung a canopy of dense black smoke, lighted up now and then by the bursting shells from the numerous arsenals scattered throughout the city."

Restoring Order

Martial law was proclaimed under Brigadier General Shepley, whose first task was to control the fire and the looters and rioters. Historian Rembert W. Patrick states, "Bucket brigades threw water on unfired buildings...wooden walls were battered down...untouched buildings were leveled by explosives. Within four hours...the fire was contained." By late afternoon, the army had suppressed the looters. That night, guards were posted at each corner to assure safety for the residents.

"I would extend leniency…return to your allegiance, renew your support to the government and become good citizens."

—Vice-President Andrew Johnson

OPPOSITE *Residents and retreating Confederate soldiers flee over the James River Bridge. Before abandoning the city soldiers blew up the arsenal, igniting hundreds of rounds of shells. Wooden ships loaded with explosives erupted, setting the bridges on fire.*

RIGHT *Union soldiers take possession of the burning city. Martial law was proclaimed under Brigadier General Shepley, whose first task was to control the looters and rioters, as well as the fire.*

Historical Background

The Civil War was drawing to a close. The Battle of Petersburg took place outside Richmond on Saturday, April 1, and ended on Sunday, April 2 with General Robert E. Lee's line of defense broken and the Confederate troops evacuating towards Appomattox. Lee telegraphed, warning residents to abandon the capital of the Confederacy. The news spread quickly, and the entire wartime population of nearly 100,000 rushed to gather up belongings. Banks were opened, despite being Sunday, so people could withdraw deposits and empty safe deposit boxes of hoarded stocks, deeds and plate. Millions of dollars in Confederate bonds and currency were tossed into the streets, and the archives of the Confederacy were packed and put on railroad cars. President Davis and his cabinet left on a train bound for Danville at about 9 p.m. The military controlling the city determined that no supplies that could aid the Union should remain.

Lincoln's Visit to the Destroyed City

The fire had destroyed almost twenty blocks of the business district, including hotels, banks and government buildings. Relief supplies were distributed to those in need of food. The army cleared debris from the streets in preparation for a visit by President Lincoln on April 4th. He arrived by boat, and walked from the river up Main Street to the home of Jefferson Davis, the former White House of the Confederacy. That afternoon he rode in an open carriage through the city. Ten days later he would be assassinated at Ford's Theater.

Crowds rushed into the streets, surrounded by roaring flames. Entire blocks were on fire, and thick black smoke rose into the air. Drunken looters appeared, and fought with fleeing soldiers. The fire spread on the wind, from the warehouses to the mansions of the rich.

At 8:15 a.m. Mayor Peter H. Mayo, with a delegation of city officials, rode out to meet Major Stevens, who was in charge of Union troops. He offered a note of surrender, and at 10 a.m. he formally surrendered the city of Richmond, on the steps of the Virginia capitol, to General Weitzel. The Union cavalry rode up Main Street during the height of the fire.

Sultana Steamship

DATE · APRIL 27, 1865
DAMAGE · 1,800 DEAD
CAUSE · FAULTY BOILER

Historical Background

On March 20, 1865 Confederate authorities sent a large group of joyous Union soldiers to Camp Fisk (or Four Mile Camp) at Vicksburg, Mississippi. Over 5,500 prisoners from Southern prisoner of war camps, such as Andersonville in Georgia and Cahaba prison near Selma, Alabama, were to be sent home. Many prisoners weighed as little as 80 pounds due to the wretched, unsanitary conditions in which they had been held.

The steamer Sultana docked on April 23 in Vicksburg, with a bulging seam on one of her boilers, which was reluctantly patched by R. G. Taylor, a local boilermaker. The military was offering five dollars per enlisted man and ten dollars per officer for transport north up the Mississippi. The owner of the Sultana, Captain Mason, needed the money, and entered into a bargain with Captain George Williams of the U.S. Army, in which all of the prisoners would be transported on the Sultana, rather than split among a number of other boats. The Sultana left Vicksburg on April 24 with nearly 2,400 people packed on the decks "like hogs." Despite the overcrowding, the prisoners onboard were jubilant to be going home. There was singing and stories of what they would do when they reached home. The boat made a brief stop in Helena, Arkansas and departed Memphis, Tennessee at 11 p.m. on the 26th.

At 2 a.m. on April 27, just three days into the trip, three of the four boilers exploded with such fury that people asleep onshore were awakened.

The top decks collapsed onto the lower decks, trapping men under the timbers. Hot coals ignited the

BELOW *The explosion tore through the Sultana directly above the boilers, flinging live coal and splintered timber into the night sky like fireworks. Whole chunks of the upper decks, including most of the pilothouse, were blown clear of the boat.*

ABOVE *The first steam fire engine built in South Carolina. Construction started in 1862. This picture shows it finished and in working condition in 1866.*

The Rescues

Just before dawn, the Sultana came to rest at the head of Chicken Island and the citizens of Memphis, finally aware of the wreck, put out in numerous small boats and rafts to rescue those who were still alive, clinging to trees and bushes. Cutters from the U.S.S. Grossbeak, the USS Essex and the USS Tyler helped in the rescue efforts, pulling hundreds out of the river.

Survivors

Miraculously, many of the men survived and watched as the night sky was lit up by the flames of the Sultana. There were numerous stories of soldiers clinging to boards and even dead horses.

The Investigation

Investigations afterwards raised suspicions that the Sultana captain had bribed Captain Williams to transport all the released prisoners. The investigators reached the conclusion that the explosion was caused by an insufficient supply of water in the boilers, inducing them to overheat. The Secretary of War, Edwin Stanton, exonerated one Union officer, Captain Frederick Speed, and the Army closed the book on the Sultana.

splintered decks, and survivors described the agonizing shrieks and the stench of burning flesh of hundreds trapped on the boat. The water from the boiler added an additional terrifying hazard: scalding steam and water showered hundreds of prisoners.

J. Walter Elliott saw "mangled, scalded human forms heaped and piled amid the burning debris on the lower deck." The steamboat was still moving forward, and the wind blew the fire toward the stern of the vessel. Roaring flames enveloped the boat within 20 minutes of the explosion. Many men fought their way to the bow of the boat in hopes of saving themselves. However, the two paddlewheels fell off and the boat spun in the water, propelling flames toward the bow. Hundreds of men were forced to jump into the icy river, which was soon filled with men drowning and fighting with each other for survival. John L. Walker later wrote, "While under water I was driven back two different times by others jumping on me. I could feel arms and legs and bodies of men all around me."

Approximately 800 survivors were taken to the hospital or the Soldier's Home, although many of them died

HISTORICAL ACCOUNT

Historian Jerry O. Potter describes the scene: "The explosion tore instantly through the decks directly above the boilers, flinging live coal and splintered timber into the night sky like fireworks...Whole chunks of the upper decks, including most of the pilothouse, were blown clear of the boat."

"After all their suffering in Southern prisons getting safely within our lines, on our route homeward, congratulating ourselves on the good news and the time we were to have at home, all this, and to have this terrible calamity, hurling so many into eternity, it makes me shudder as I write.

—Arthur A. Jones, survivor, in a letter to his brother

within a few days. No one was ever quite certain how many men had been aboard the Sultana, so we are not certain how many died, although unofficial estimates have gone as high as 1,800. For several weeks afterwards bloated bodies were still being pulled from the river. Most of the bodies are buried at either the Elmwood Cemetery or the National Cemetery in Memphis.

The Ohio survivors formed the Sultana Association, which for years struggled in vain to get Congress to erect a monument to the tragedy. The East Tennessee Survivors Association erected their own monument in the Mount Olive Cemetery near Knoxville. The country soon forgot what happened, but the survivors never could. The Sultana wreckage was never salvaged, and was covered by farmland when the Mississippi shifted its course.

Barnum's American Museum, New York City

DATE • JULY 13, 1865
DAMAGE • OVER $1,000,000
CAUSE • DEFECTIVE FURNACE

The Museum Before the Fire

P. T. Barnum opened his American Museum in 1841, at Broadway and Ann Street in New York City. Flags and banners flew from the roof of the decorated building, in which he presented performers such as the midget General Tom Thumb (Charles Stratton) and "The Swedish Nightingale" soprano Jenny Lind. The museum was a popular family attraction during the 1850s, and featured caged wild animals, as well as scientific specimens, curiosities of the natural world and other oddities, such as the largest elephant ever known, Egyptian mummies and a working steam engine made completely of glass. *The New York Times* stated, "No public institution in the country pretended even to rival the geological collections of the museum either in extent or value."

The fire originated in a defective furnace in the cellar under a restaurant, and was discovered at 12:35 p.m. The alarm was sounded, and patrons and employees, realizing the imminent danger, immediately attempted to save exhibits, such as the collection of wax figures of famous notables, including Napoleon, Queen Victoria, Jefferson Davis, and Christ and his Disciples. One firefighter saved the wax figures of Tom Thumb, his wife and baby.

By 1 p.m. the fire was out of control, and a half hour later the crowd, which had swelled to hundreds, heard someone shout, "The snakes are loose" in a mistaken belief that the cage containing a live 30–foot boa constrictor, which ate whole sheep, had been shattered by the heat. At the same time a trumpet blast caused someone to shout,

"I have resolved to erect immediately in this city a museum which will be an ornament to our great metropolis."
—P. T. Barnum after the fire

"The elephant is coming!" In just two minutes the panic of the crowd, rather than the fire, caused over one hundred injuries, including some who were pushed through the windows of the neighboring Astor House Hotel.

The panic increased when the crowds heard shrieks and bellows, as the live animals in *The Happy Family* were burned to death while the whales, hippopotami and fish were boiled to death in their tanks.

At 1:30 p.m. the whole wall on the Ann Street side fell in, at 1:45 p.m. the Broadway front fell in three different sections, and about five minutes later the Park Row side fell, which spread the fire into neighboring

---

BELOW LEFT *Firefighters rescue animals from the fire. Several men are shown pulling on a roped giraffe, one man tries to hold onto a stork and a pair in the foreground guide a baby elephant past a small cage of pigs.*

BELOW RIGHT *A steam pumper (with running company in 1863) is in the Brooklyn Navy Yard for repairs.*

ABOVE *One of the museum walls collapsing into the street. Miraculously, no human lives were lost.*

The Happy Family

The Happy Family was a collection of monkeys, dogs, porcupines, guinea pigs, chickens, rats, cats and pigeons, all in one huge cage, which was to demonstrate that God's family could live in harmony.

buildings on Fulton Street. Merchants closed their stores and jammed the streets with carts, attempting to save their goods, crowding fire-fighting efforts.

By 2:30 p.m. firefighters had managed to contain the fire by relentlessly dousing the flames from both the street and the roofs of neighboring buildings. By 3 p.m.

the fire was wholly out. Barnum estimated losses at $1,000,000. The next day thousands of people visited the site, and police had to keep the crowds at bay. Another museum was constructed at Broadway and Spring Street, but it too burned in 1868. In April 1871 Barnum opened the first three-ring circus in Brooklyn.

Portland, Maine

DATE · JULY 4, 1866

DAMAGE · 2 DEAD; 12,000
HOMELESS;
$10,000,000

CAUSE · FIRECRACKER

"I have been in Portland since the fire.
Desolation! Desolation! Desolation!"
—Henry Wadsworth Longfellow

The largest fire in the United States to this date was the "Great Fire" of Portland, which began in the late afternoon during the 4th of July festivities. A boy threw a firecracker into a boat-builders' shop adjoining a lumberyard on Commercial Street, which started the blaze in a pile of wood shavings. Gale strength winds blowing from the south soon ignited the wooden houses nearby. Blazing embers fell on the roofs of nearby warehouses. By the time the fire department reached Brown's Sugar House, the fire was out of control. The few steam engines and hand pumps available soon exhausted the water supply from wells and cisterns, and firefighters, coming from as far as Boston and Augusta, could do nothing. The wind, strengthened by updrafts from the heat, became a hurricane of fire, carrying blazing debris through the air. By nightfall the fire could be seen for miles, and illuminated the sky with huge sheets of flame.

The fire swept diagonally across the city, from Commercial Street to Back Cove, and up to Monjoy Hill. It raged throughout the night, and only ended when there was nothing else to burn. Tents were set up on Monjoy Hill, and relief supplies from Boston arrived immediately.

The reconstruction prompted one reporter from the *Boston Journal* to write, "The fire has put Portland fifty years *ahead*."

HISTORICAL ACCOUNT

Historian Augustus F. Moulton wrote, "The rapidity of the progress of the conflagration was beyond conception. Structures of wood, brick and stone went down with roar and rattle of falling walls in a few minutes...the iron rails of the horse car tracks in the center of Middle Street were warped and twisted out of shape by the heat, and stone and marble were cracked and crumbled."

The Rebuilding

Rebuilding began as soon as the charred ruins could be cleared. Led by a strong mayor, Augustus E. Stevens, streets were widened and straightened, Lincoln Park was established, and an adequate water supply was ensured, when pipes were laid to bring water from Sebago Lake.

The Damage

The burnt district was about 300 acres, and included every bank and newspaper office, every book store, eight churches, eight hotels, the new city hall, the Post Office and every lawyer's office. Some 1,800 buildings were destroyed, and over 10,000 people were left homeless. The Mayor sent a plea to the Mayor of Boston: "Thousands of our people are homeless and hungry. Can you send some bread and cooked provisions?"

Chicago, Illinois

DATE · OCTOBER 8, 1871
DAMAGE · OVER 300 DEAD;
BURNED OVER
2,024 ACRES; LOSS
OF $196,000,000
CAUSE · UNKNOWN

"No city can equal now the ruins of Chicago, not even Pompeii."

—J. W. Goodspeed

Who has not heard the story that Mrs. O'Leary's cow kicked over a lamp in her barn and started the "Great Fire" of Chicago? Interestingly enough, some of it seems to be true.

The weather had been excessively warm that autumn, and the city was very dry. There had been only a couple inches of rain since the 4th of July. On Saturday October 7, the day before the Great Fire, there had been another fire, which burned four blocks, and exhausted many of the firefighters.

A small fire began at about 9 p.m. on Sunday, October 8 behind Patrick O'Leary's house in the Western District. The exact particulars of the fire's origins are unknown. The working class neighborhood had wooden houses packed close together. The exhausted firefighters were slow to respond, and by the time they arrived it was too late. A strong wind from the southwest was blowing the flames straight towards the downtown district, with its beautiful but highly flammable wooden sidewalks. By 10:30 the fire was reported to be out of control, and sheets of flame quickly burned through the Western District. It was estimated that 65 acres of the city burned every hour during the course of the blaze, feeding on the wood sidewalks, streets and buildings.

By midnight firebrands whipped by the winds had leaped across the South Branch of the Chicago River, falling on the Adams Street Bridge, and began to rain on the business district. Fire-bells all over the city were ringing out their warnings, and the sleeping populace awoke to find themselves in great danger.

By 1:30 a.m. the fire had reached the Courthouse and the Chamber of Commerce. Half-clad men, women and children were dragging whatever they could carry through the streets in search of safety.

Howling animals ran loose, adding to the confusion of the refugees as they struggled towards the North Division.

By 1:30 the fire began to burn the State Street Bridge, and the flames moved quickly into the North District. Panicked people crowded the Rush Street and Randolph Street bridges, moving northward. Husbands were separated from wives, and children wandered alone.

By dawn on Monday the 9th the fire was burning the mansions of Chicago's wealthiest citizens, such as McCormick, Kinzie and Rumsey on the North Side. Fire burned the Waterworks, leaving only the gothic stone Water Tower standing. The flames continued advancing to North Avenue, the northern border of Chicago.

The prayers of the people were answered when rain started falling late Monday night and continued into

The City Before the Fire

Following the Civil War, Chicago had grown to become the center of manufacturing and agriculture for the entire West, with ten railroads shipping commercial traffic in wheat, corn, hogs and cattle throughout the country. The city had a population of over 300,000 people, and spread over 23,000 acres. It also had over 50 miles of wood streets, over 500 miles of wooden sidewalks and mile after mile of wooden houses stretching west.

BELOW *Bird's-eye-view of Chicago as it was before the Great Fire.*

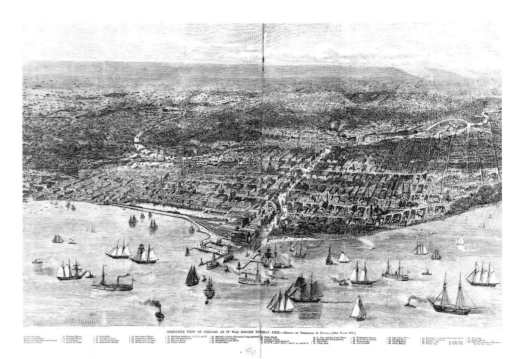

According to an account by Bessie Helmer, "It was like a snowstorm only the flakes were red instead of white."

ABOVE *Chicago in Flames: scene at Randolph Street Bridge. Half-clad men, women and children were dragging whatever they could carry through the streets, in search of safety.*

RIGHT *Weary and undermanned firefighters made their best, but ultimately futile, efforts to stop the giant fire.*

The Damage

The fire burned the Grand Pacific Hotel on La Salle Street, and moved up the block, burning stores, banks and office buildings. It swept up State Street, and burned the magnificent Field & Leiter's Department Store and the Palmer House hotel. At the height of the fire, there were 500 buildings engulfed in fire at the same time.

Tuesday. According to J. W. Goodspeed's account of the fire, "Women broke into wild sobs of relief and strong men embraced and burst into tears of thanksgiving."

The exhausted refugees camped out in the cemetery, which would become Lincoln Park, or on the prairie to the west of the city. Chicagoans of all social strata mingled with each other in their desperate plight.

Troops, led by General Philip Sheridan, patrolled the streets, guarded the relief stations and brought a degree of calm to the city. The Relief and Aid Society divided the city into sections, and oversaw the distribution of donated materials, which also included medicines and materials for building. The entire nation responded with millions of dollars for relief, and Chicagoans soon

---

BELOW *The scene at the Randolph Street Bridge after the fire. The "Great Rebuilding" lasted three years, and by 1873 Chicago could claim to be the premier city of the West.*

Restoring Order

Mayor Mason and the City Council established a Relief Committee whose task was to provide food and water. They also asked General Philip Sheridan to take command of the city, and martial law was imposed. Anxious citizens returned to find the city completely wrecked. The burnt district was four miles long and three-quarters of a mile wide. Approximately 18,000 buildings and 8 bridges had been destroyed and 100,000 people were homeless. The fatalities were surprisingly low, considering the extent of the conflagration, with estimates just over 300 persons.

began to rebuild in the ruins. A third political party, the "Fireproof Party," attempted to run candidates in the November election who would improve building codes, but was doomed to failure, running against the rich and powerful city leaders.

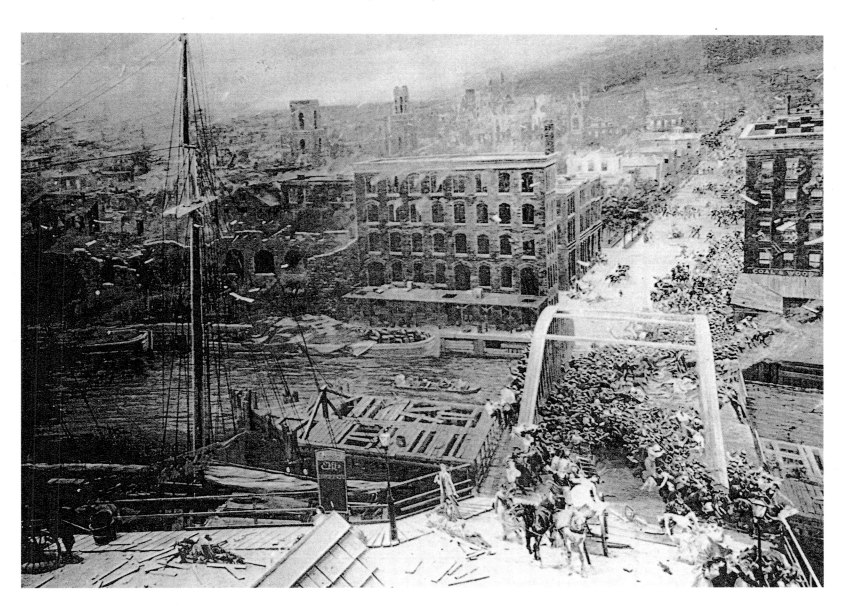

Peshtigo, Wisconsin

DATE · OCTOBER 8, 1871
DAMAGE · 1,200 DEAD
CAUSE · LIGHTNING STRIKE

HISTORICAL ACCOUNT

J. W. Goodspeed reflected, "We saw many families who had been kept in the water all night...a large number were drowned by the cattle and horses that, maddened by the fire, rushed into the water." One woman asked Father Pernin, "Is this the end of the world?"

America's most disastrous forest fire to date took place a few hundred miles to the north in Wisconsin on the same day as the "Great Chicago Fire."

The fall of 1871 was so dry that the swamps had dried up. It had not rained since July 8 and the cranberry bogs had become dry as peat. The pine trees were tinder ready for

Fire Behavior

Firefighters of today would rate Peshtigo's fire as extreme fire behavior, characterized by high, sustained rate of fire spread, rapid buildup of intensity and blowup situations, with fire whirlwinds and horizontal flame sheeting.

burning. Wildfires had been started by lightning strikes and were burning unimpeded through the forests.

By October 4 the smoke was so thick on Green Bay that ships blew their foghorns in daytime. Woods on both sides of the bay were burning. By the 7th farm families who stayed to fight the fire found themselves trapped.

On Sunday the 8th residents of Peshtigo could feel the hot winds increasing, and the sky was dark with smoke. At about 9 p.m. a low moaning roar could be heard from the southwest. The roaring increased, and suddenly a whirling fire swept down and landed on the sawdust-covered street in front of the Peshtigo Company's rooming house. The wind, fueled by the updraft of the fire, lifted the roofs off houses, toppled chimneys, and sent hot embers raining fire on the town. The pine sidewalks caught fire, and houses flared up instantly. The intensity of the flames consumed available oxygen, and many people asphyxiated and fell where they stood.

The worst of the firestorm, which some compared to a hurricane of fire, lasted about an hour. In that short time the entire village was wiped out.

Survivors straggled out of the river around dawn. Refugees walked to Marinette, in search of food, and told the story of the fire. A steamboat captain got to Green Bay with the news, and telegraphed Governor Fairchild and various mayors. It was a full twenty-four hours before the world knew of the fire at Peshtigo. Even then it was the Chicago fire which claimed the majority of the headlines, even though the Peshtigo fire claimed over three times the number of lives. A Peshtigo survivor said, "The names of half the dead will never be known."

Survivors coming back into town found it had disappeared completely. Bodies were horribly burned, and in some cases were just heaps of ash that the wind carried away. Entire families had been wiped out. Sometimes there was only a metal pin or button to show that a person had died. Six days after the fire, the last body was buried. He was a foreman of a logging camp.

Most of the early relief came from neighboring towns. Food and clothing arrived from Madison by rail, as quickly as it could be gathered. Milwaukee sent blankets, quilts and bedding. The Peshtigo Company decided it would rebuild, and other timber industries followed their lead. Within a remarkably short time the town was rising from the ashes. The devastation of the fire had also created opportunities: clearing the land of trees and creating thousands of tillable acres. Wool, butter and cheese factories replaced the sawmills.

Franklin Tilton wrote, "Within a half hour and some say ten minutes, of the time the first building caught fire, the entire village was in flames…some were burned to death within a few feet of the river, some in their houses and some on the roads trying to escape." Father Pernin, the priest, thought of the drunken lumberjacks passed out in the sleeping quarters above the bars and reflected later "it was impossible to know how many of them might have survived if they had been sober." The roads were filled with panicked people, animals and wagons heading towards the Peshtigo River. "There was tragic confusion when the frantic, struggling mass of people tried to escape from the fire. Some were trampled underfoot."

The City Before the Fire

Peshtigo is situated about 49 miles northeast of Green Bay, on the Peshtigo River. First settled in the 1840s, following the Civil War, it became a boomtown with a population of around 2000 permanent residents. Peshtigo had a doctor, a general store, a meat market, a hardware store and a piano dealer. It also had the Peshtigo Co. sawmill, which ran 97 saws, and had a daily cut of 150,000 feet of lumber. Sawmills and lumbering dominated life. The houses and roofs were lumber, the sidewalks were made of wood, the streets had sawdust on them, and even the mattresses were stuffed with sawdust.

"Once in water up to our necks, I thought we would, at last be safe from the fire, but it was not so. The flames darted over the river as they did over the land…the air itself was on fire."
—Father Peter Pernin, survivor

The Destruction

The fire burned 16 other towns and 1.2 million acres. It had killed over 800 in Peshtigo alone and 1,200 in the whole area.

Writing in the *Boston Journal*, Edward King described the scene: "Approaching the burned district toward noon [the next day], one might readily have fancied himself in a recently captured and bombarded town. Boston's center seemed suddenly to have vanished."

Boston, Massachusetts

DATE · NOVEMBER 9–10, 1872
DAMAGE · 776 BUILDINGS, 65 ACRES; $75,000,000; 14 KILLED; INCLUDING 11 FIREFIGHTERS
CAUSE · STEAM ENGINE

Saturday evening, November 9, 1872 was very clear and calm. Stores had closed, and many people had traveled to their homes just outside the city. A spark from a small steam engine, which operated an elevator in a dry-goods wholesale warehouse on Summer and Kingston Streets, started the fire. It was discovered at 7 p.m., but the first alarm was not given until 7:24. Had there been modern fire alarms, which automatically transmit a fire emergency to the station, it's possible firefighters could have arrived at the scene earlier, and prevented the flames from spreading.

ABOVE *Boston firefighters shown with their steamer on Federal Street, with the remains of the Post Office in the background. The firefighters could not position their equipment in the narrow streets, and a lack of water pressure prevented them from reaching beyond the second or third floors of buildings.*

In an odd twist of fate, most of the horses in Boston were suffering from a plague-like disease, which prevented them from standing without support. Many men who could have helped fight the fire had been hired to haul horse-aid equipment, leaving few people to respond to the fire.

By the time firefighters arrived the fire had shot up the elevator, and the building fire was out of control. Flames soon leaped to Otis Street, then onto "Beebe's Block" of dry-goods stores at Winthrop Square, "con-

"Boston—we have seen in thy ash heaps and shattered facades more beauty than wealth can purchase, wrecks of buildings but no wrecks of men, ruined storehouses but no ruined intellects. We are all proud of thee today."

—newspaper writer

suming great quantities of hoop skirts, hats, caps, woolens, boots and shoes."

Chief Engineer John C. Damrell issued a general alarm for engines and firefighters from all over the city and shortly afterwards he called for engines from Cambridge and Charlestown. The firefighters could not properly position their equipment to effectively fight the fire because of the narrow streets, and a lack of water pressure prevented the water from the fire hoses from reaching beyond the second or third floor.

Stone facades cracked and exploded in the intense heat, probably exceeding 1,000 degrees and the flames spread from building to building, feeding on the wooden sills, cornices and wooden mansard roofs. Roofs fell

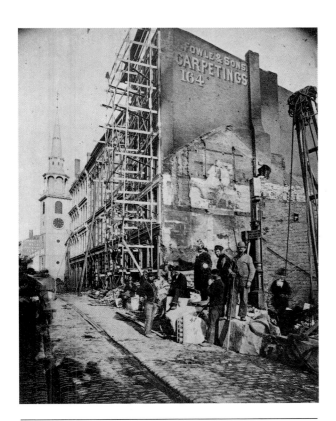

ABOVE *Ruins of the Great Fire, looking toward the harbor. The fire consumed 65 acres of the city.*

The Mansard Roof

Mansard roofs originated in France during the reign of Napoleon III. Part of the Second Empire style, they featured wood sloped roofs which formed the upper story. They were popular in America both on houses and on commercial buildings. The old Bates mansion in the film *Psycho* is a good example of a mansard roof. While attractive, they are very flammable.

The Investigation and Reaction

The investigation afterwards placed some of the blame on the construction of the wooden mansard roofs. Stricter building codes were put in effect, and fireproof materials were mandated.

in with a crash almost every minute, carrying with them the floors beneath.

The fire spread northward towards State Street and escaping gas from buildings on Washington Street exploded. Soon there was a wall of fire from Washington Street to the bay. Banks, offices, stores and churches were totally destroyed. As the night went on, the scenes became more gruesome—firefighters in long rubber coats dragged hoses along flooded streets, while flaming fabrics fell on the crowds of seamstresses trying to escape.

By midnight the fire had spread to Congress Street, and by 1 a.m. it had crossed Pearl Street. U.S. Marines were brought in to assist with fire-fighting, and at 2 a.m. the drastic decision was made by Mayor Gaston and Chief Damrell to create firebreaks by dynamiting buildings. Unfortunately, the gunpowder was poorly used and only helped spread the fire.

The fire burned out of control all day on Sunday, raging through the business district. Gas had been turned off in the entire city earlier that day, but late that night exploding gas in the sewers ignited additional buildings.

The fire was finally brought under control early Monday morning through the continuing efforts of the firefighters, and because there was little left to burn.

In the week following the fire thousands of laborers were clearing the rubbish, marking the streets and recovering property such as safes. Large work crews were tearing down dangerous masonry walls, while smoke and steam continued to come up out of the ruins.

Help began to arrive immediately and monetary subscriptions for rebuilding and supplies were raised in many cities including New York, Chicago, Pittsburgh, St. Louis and Baltimore. Even London, Paris and Liverpool sent monetary contributions.

Conway Theater, Brooklyn, New York

DATE · DECEMBER 5, 1876

DAMAGE · 295 DEAD

CAUSE · GAS LIGHT IGNITING SCENERY

The Conway Theater at 313 Washington Street, Brooklyn, was the site of the worst theater fire in the United States up to that date. It was later surpassed by the Iroquois Theater fire of 1903 in Chicago.

The last act of *The Two Orphans*, starring Miss Kate Claxton, was reaching a climax when a gas light from the overhead flies fell onto a stretched canvas, which represented a house roof. The scenery burst into a sheet of flame almost instantly, and the auditorium was filled with smoke. A panic ensued, and people rushed for the doors.

Within three minutes the fire reached the roof, and a massive wall of flame trapped the patrons on the third tier, which was crowded with 400 young men and boys. They fled to the only stairway, which was narrow and twisted, and in their panic pushed each other down the stairs and were trampled by others from behind. The staircase collapsed from the accumulated weight, and plunged them onto the lobby floor, which in turn gave way and fell into the flaming cellar. The mass of men was entangled in the twisted wreckage, and many were burned beyond recognition.

In less than 20 minutes, the mansard roof collapsed into the theater, followed by the collapse of the Johnson Street wall, covering the street with debris. The gas, which had not been turned off, burned fiercely, lighting

The Victims

The *New York Daily Tribune* described the charred and blackened corpses: "Some of the bodies were stiffened in almost a sitting position...arms were flexed and hands clenched in the act of pushing...knees were bent and legs drawn up as though fighting off an advancing foe." Identification of the victims depended mostly on watches or other items such as rings. Hundreds walked tearfully down the rows of victims searching for sons and brothers. Funerals of identified bodies began on December 8th and continued for two days. Mayor Schroeder of Brooklyn declared that the city would bury the 100 unidentified bodies together in a common grave in Greenwood Cemetery.

LEFT *The actors tried to calm the audience when fire broke out in Brooklyn's Conway Theatre, but to no avail.*

ABOVE *Audience members struggled down the staircase, which collapsed from the accumulated weight and plunged them onto the lobby floor, which in turn gave way and fell into the flaming cellar.*

up the scene and sending a cloud of white steam into the air. A few minutes later the eastern wall collapsed, and the Brooklyn theater was a smoldering ruin.

Hardly an hour had elapsed from the time the fire was discovered until the building was totally demol-

A Hero

The bravest man at the fire was Charles B. Farley, the assistant engineer of the fire department, who was first on the scene. He ran into the burning building, and singlehandedly pulled over 40 people out to the street. Many had been bruised and beaten, but survived. He and an assistant went to the roof of Dieter's restaurant next door and ordered Engines No. 5, 7 and 8 to turn their streams of water on the neighboring houses in order to save them.

ished. Streams of water were kept pouring on the ruins all night, and at daylight Mayor Schroeder directed the men to begin the search for the missing.

Firefighters and police were able to enter the building, while a dense crowd filled Washington and Johnson Streets. Bodies were removed from the horrible pit, covered with blankets and placed on carts bound for the three temporary morgues.

Despite a frigid snowstorm, public services to honor the dead were held at the Brooklyn Academy of Music and the funeral procession proceeded down 6th Avenue, watched by thousands of mourners. The German Society sang Kuliak's *Abendlied*, beginning "Under the greenwood there is peace."

RIGHT *Fire rages through the mansard-roofed theater while crowds gather outside. Firefighters battle the flames from the adjoining rooftops.*

BELOW *Family members identify victims. Almost 300 people died in the blaze.*

The Reaction

The fire prompted an international outcry for stricter safety standards at theaters and more frequent fire inspections. Laws were passed requiring more exits, fire escapes and fireproof fabrics.

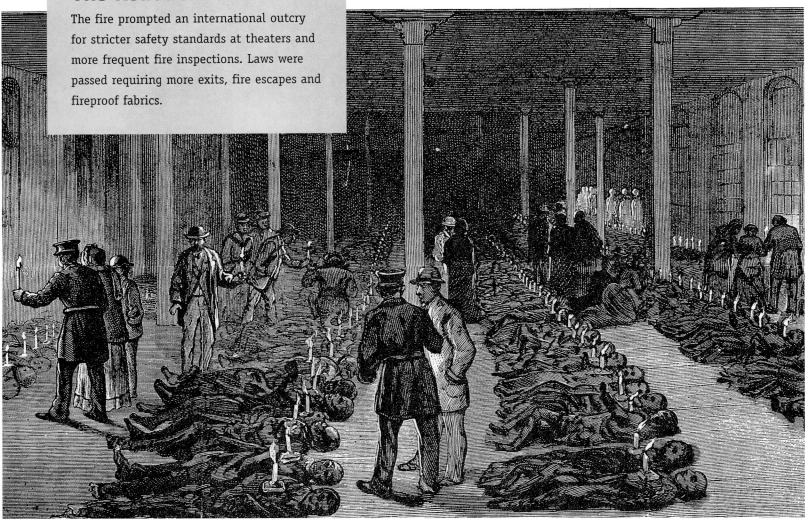

Ring Theatre, Vienna

DATE · DECEMBER 8, 1881

DAMAGE · 794 KILLED

CAUSE · GAS LAMP IGNITING STAGE SCENERY

At about 7 p.m. the glamorous crowd, which numbered nearly 2,000, was still searching for their seats while last minute activities were underway backstage. Some of the actors had already taken their places on stage and the overture was about to begin when a lamp fell on the stage igniting the scenery. The flies and network of rope, canvas and wood above the stage burned like dry tinder. Five water taps located above the stage were never used because the firefighter posted to the back-stage area had fled. Frightened actors and stage hands threw open a door to the Hohenstanfengasse, and the stage curtain blew out to the audience like a fiery

ABOVE This drawing captures the moment inside the theater when actors and stage hands opened a door backstage, drawing in air and blowing the stage curtain towards the audience, setting fire to the rest of the auditorium.

tongue, setting fire to the auditorium.

An iron stage curtain, which also might have prevented the spread of the fire to the galleries, was never lowered because the workman had fled from the flames. A fast-thinking worker turned off the gaslights at the source,

The Theater Before the Fire

The Ring Theatre was built in 1873 to meet the entertainment demands of the people attending that year's Great Exhibition. Offenbach's opera *The Tales of Hoffmann* was about to be performed for the second time in Vienna.

The Investigation

American Architect and Building News covered the trial and reported that gross neglect was charged against the management of the theater, as well as city inspectors and police officials. The law prescribed the use of oil lamps in the corridors of theaters, so that the exits could be readily seen in the event of the failure of the gas. This precaution was not used, although it is unclear as to how many people it would have saved.

The concluding words of the indictment stated, "Except the few fortunate persons who succeeded in saving themselves or their relatives, no one was saved, either by the police or by the fire brigade." The fire consumed the theater too quickly.

> "Nothing like this horror has ever occurred before in this city."
> —*Wiener Allegemeine* newspaper

LEFT *Firefighters stretch a blanket to catch a woman who leaped from the second story—a height of over 60 feet. Hundreds were saved by this method.*

The Dangers of Gas Lighting

Before the invention of the incandescent electric light, by Thomas Edison in 1879, all buildings were lit by gaslights. Like today's gas kitchen stoves, gas was run through pipes into fancy lighting fixtures. Adjusting a twist knob could turn the flame up or down, thus changing the light. Because of the open flame, there were hundreds of fires every year. Anything highly combustible, like fabric, paper or clothing, might suddenly catch fire if it brushed into the flame. Theaters used gaslights both onstage and in the auditorium and vestibules. Theaters usually kept "fire wardens" posted throughout, with buckets of water to extinguish a fire, and they also had iron curtains to lower between the stage and the audience to stop a fire.

for fear of an explosion, which plunged the theater into total darkness. The flames shot up to the dome in an instant, trapping the spectators, and then burst through the roof, engulfing the entire building within minutes.

When firefighters arrived, not even ten minutes after the start of the fire, it was so widespread that they found it impossible to move beyond the first tier of the theater, due to the suffocating smoke.

The fire was out by dawn, although the left staircase re-ignited at about 1 p.m., and a small fire in the ruins broke out again the following evening.

Newhall House, Milwaukee, Wisconsin

DATE · JANUARY 10, 1883
DAMAGE · ESTIMATED 75 DEAD
CAUSE · ELEVATOR MOTOR MALFUNCTION

The fire started at the rear of the hotel, seemingly caused by a malfunction in the elevator engine just before 4 a.m., during a light snowfall. It was still dark when the alarm sounded, and Fire Chief Lippert quickly arrived at the hotel. At 4:19 a.m. the whole fire department was called to the scene because the entire south front of the hotel was a mass of flames. The fire trapped screaming guests on the upper floors as it spread up the open stairways. Many people leaped from the windows of the upper stories, and were killed or injured when the canvas held by the firefighters to catch jumpers was repeatedly pulled from their hands. One of the greatest obstacles to saving people was the network of telegraph wires. They surrounded the building and made it impossible to hold a canvas in such a position that the people leaping would not hit the wires first. There were frequent shouts of "Cut down the wires, chop down the poles," although nothing was done.

BELOW *The Newhall House was constructed entirely of wood, had 300 rooms and the façade had been covered with a brick veneer. It was correctly considered by many guests to be a firetrap.*

The Investigation

George Scheller, the owner of the Newhall Bar, was arrested on January 16 for setting the fire. Precautions were taken by police to prevent a lynching. On February 28 the Grand Jury found him innocent, and attributed the fire to a malfunctioning elevator.

The Building Before the Fire

The Newhall House hotel was built in 1857, on the corner of E. Michigan and N. Broadway, by Daniel Newhall, at a cost of $270,000. It was six stories high, had 300 rooms and at the time it was built, was considered the finest hotel in the West. The structure, made entirely of wood, had been covered with a brick veneer making people think it was fire-proof, when in reality it was merely a wooden building.

A Famous Survivor

General Tom Thumb, a midget made famous at P. T. Barnum's museum of curiosities, and his wife Lavinia were guests at the hotel at the time of the fire. *The New York Times* reported, "They were rescued by Police Officer O'Brien who took one of the little people under each arm and carried them downstairs and across the street to the American Express Company." It was also noted that Mrs. Thumb "bravely began relieving the sufferings of those around her by supplying them with water." Later that evening the General told a reporter, "Yes, the policeman entered our room...but I went down the ladder alone."

"The building was 26 years old and known to everybody as a fire-trap… there was no insurance company in Milwaukee that would take a risk on it."

—Larkin B. Day, Milwaukee eyewitness

At 5 a.m., an hour after the fire started, the upper floors of the hotel collapsed in a mass of flames, carrying the lower floors into the cellar. The crash could be heard for blocks. At 5:30 the unsupported Broadway front gave out and crashed to the street, followed by the remaining facades.

By 6 a.m., firefighters had extinguished the flames, and the hotel was nothing but a mound of smoldering wood and brick.

The fire prompted revisions of the hotel building laws, requiring fire escapes and ropes in every room. Rooftop water tanks became required, to provide sufficient pressure for fire-fighting within hotels, and individual chemical extinguishers were mandated in hotels throughout the country.

BELOW *The fire consumed the Newhall House in just over two hours. The ruins were little more than a mound of rubble.*

Galveston, Texas

DATE · NOVEMBER 13, 1885
DAMAGE · $2,500,000; 568 HOMES; 52 BLOCKS
CAUSE · UNKNOWN

The first volunteer fire-fighting company in the city was Hook and Ladder Company Number 1, founded in 1843. By 1884, the city had a salt-water fire-fighting system, which would draw water from the bay and pump it to 110 hydrants. In 1885 a professional fire department was formed, but it only had 6 steam engines, one hook and ladder and 18 paid firefighters. It was just over a month old when the worst fire in the history of Galveston struck.

ABOVE *The smoldering ruins of Galveston after the fire. The heat became so intense that hot, scorching whirlwinds swirled through town, igniting block after block.*

The City Before the Fire

The city of Galveston is built on a barrier island in the Gulf of Mexico, which varies in width from $1\,^1/_2$–3 miles, and is 27 miles long. The city received its charter in 1839 and, following the Civil War, began to thrive as a beach resort. The colorful Beach Hotel was built in 1882, and numerous Victorian hotels and houses followed, many designed by Texas's first professional architect, Nicholas Clayton. By 1880 the population was 22,000, making it the largest city in Texas.

It began at 1:40 a.m. in a small foundry near Strand and 16th Street. A 30-mile-per-hour northerly wind was blowing, which quickly spread the flames over a large expanse of Texas pine wood buildings. A delay in sounding the general alarm, combined with an inexpe-

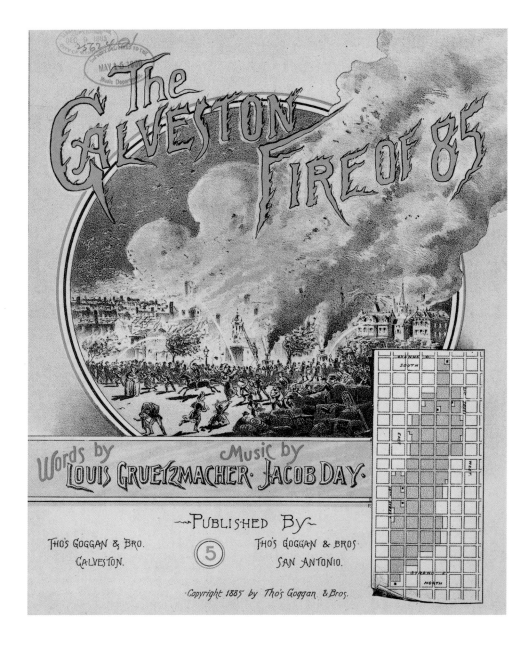

The Damage

At 6 a.m. the wind began to die down, and by 9:30 a.m. firefighters stopped the fire at last within two blocks of the beach, at Avenue O. The burnt district covered 52 blocks and included over 500 homes.

rienced fire-fighting force, allowed the fire to run unchecked. People roused from their sleep by the noise of the fire fled in terror. The heat became so intense that hot, scorching whirlwinds swirled through town, igniting block after block. By 4 a.m. the flames had covered seven blocks, and a shift in wind, which had increased to 60 miles-per-hour, veered the fire to the east, where it moved from the wooden shanties to the mansions of the wealthy.

"Please accept my deep sympathy in your misfortune. You may draw on me at sight for $5,000…to relieve the needy."

—Jay Gould, New York banker

The Rebuilding

Business resumed within a few days, and rebuilding began immediately. Ornate houses built of brick and stone replaced those that had burned, and by the 1890s life in Galveston was booming again.

ABOVE LEFT *Sheet music titled "The Galveston Fire of '85," by Jacob Day.*

LEFT *The burnt district covered 52 blocks and included over 500 homes.*

Opera Comique, Paris

DATE · MAY 25, 1887
DAMAGE · 200 DEAD
CAUSE · GAS LIGHT IGNITING
STAGE CURTAIN AND SCENERY

At 8:30 p.m. the audience was enjoying the first act of *Mignon*, when sparks from a row of gas jet lights fell on the stage, instantly igniting the curtain and scenery. In a few minutes the stage was roaring like a furnace and the auditorium was densely filled with smoke. The actors and singers fled the stage, some going back into the wings and some into the audience.

The iron drop curtain was not lowered because the person responsible had fled, so the flames spread with alarming rapidity. The gas lights went out, and occupants trapped in the upper galleries were left to grope their way out of the passages in blinding smoke and total darkness.

The fire engines did not arrive until 9 p.m., a half hour after the alarm was given. This was later attributed to crowded streets and old, hand-drawn fire-fighting equipment. By that time the building was a raging furnace. The panic-stricken throng pushed towards the five doors opening to the street, but two of them were locked. As people behind crushed those in front, a pile of bodies blocked the doors. People in the gal-

The Theater Before the Fire

The Opera Comique was erected on the Boulevard des Italiens in 1840, replacing one built in 1781. The backstage area was seven-stories high, with wooden bridges and stairways connecting the levels, while the auditorium had galleries reached by narrow, angled stairways.

The Reaction

Numerous benefits were held to aid the victims, and subscriptions came from the Conte de Paris as well as from abroad. More stringent rules were passed for theater safety, and older theaters were switched from gas to electric lighting.

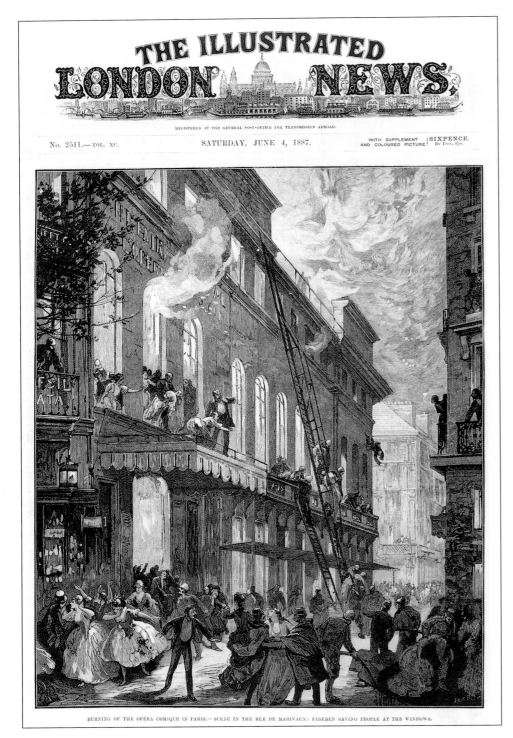

THE ILLUSTRATED LONDON NEWS.

REGISTERED AT THE GENERAL POST-OFFICE FOR TRANSMISSION ABROAD.

No. 2511.—VOL. XC.   SATURDAY, JUNE 4, 1887.   WITH SUPPLEMENT AND COLOURED PICTURE | SIXPENCE. By Post. 6½d.

BURNING OF THE OPÉRA COMIQUE IN PARIS.—SCENE IN THE RUE DE MARIVAUX: FIREMEN SAVING PEOPLE AT THE WINDOWS.

"There is no need of a great poet to raise his voice to arouse in all of us deep indignation against those persons who, not having protected our lives…do not take the trouble after the catastrophe to count the dead or protect the remains of the victims against the most horrible of profanations."

—*French* newspaper

The Victims

Crowds outside the theater were desperately searching for family members from whom they had been separated. The first bodies were brought out that night and taken to the National Library, which had been converted to a temporary morgue. By 1:30 the following afternoon 12 bodies had been recovered, but by 6:30 the number had grown to 56. Many were burned beyond recognition. Many girls were identified only by their jewelry.

LEFT *Scene in the Rue de Marivaux as firefighters carry theater patrons down ladders.*

leries leaped out windows and struggled down the twisting stairways, only to find their way blocked.

Shrieks for help from inside the theater caused firefighters and police to rush in, but they were forced back by the flames. The efforts of the firefighters were useless because the water pressure was so inadequate that the water only reached half-way up the exterior walls. By 9:30 p.m. the whole roof was in flames and shortly afterwards it fell in with a crash.

Firefighters climbed to the roofs of neighboring buildings and poured water on the theater from all sides; by 1 a.m. they had managed to extinguish the flames.

The search for bodies ended on June 1 with the discovery of two human hands and part of a skull. Ragpickers sorted through the burned ruins, but police kept watch and confiscated anything of value. In a gruesome twist, the scavengers were allowed to take the bones and ashes to be sold by the pound to refineries.

The Burial

The funeral of 22 victims took place on May 30th at Notre Dame Cathedral. Inside the cathedral the coffins were placed on black trestles in the nave. The caskets contained the remains of 10 staff of the theater, as well as 12 unidentified people. They were buried in Père-Lachaise cemetery, following a half-mile procession, which was lined by 200,000 people.

Theatre Royal, Exeter, England

DATE · SEPTEMBER 4, 1887
DAMAGES: 187 DEAD
CAUSE · GAS LIGHT IGNITING SCENERY

Over 800 patrons filled the house on the evening of September 4 for the performance of *Romany Rye*. Everything was fine until the end of the third act, at 10:30 p.m., when a burning piece of flying scenery, which had been ignited by the gaslights, dropped to the stage and immediately caused the whole stage to ignite.

The fire brigade arrived five minutes after the alarm, but their efforts were useless. The actors and actresses were taken by ladder out the back windows, but the audience in the gallery was not as fortunate. The gas was turned out, which plunged the theater into darkness.

There were about 200 people in the gallery, and they rushed towards the stairway. Halfway down the stair-

well there was a sharp angle. People were pushed down and jammed into the angle, thus blocking the exit. The stairs were literally packed with bodies piled on top of each other. It was later determined that a ticket box

The Building Before the Fire

The Theatre Royal had been built in 1886 to the highest fire standards of the time. The theater was equipped with a fire hose onstage, as well as buckets of water on each level. It had numerous exits from the balconies, but they were fastened shut and could not be opened. It covered 8,500 sq. ft. in area, and was lit by gas and decorated with plaster ornamentation.

ABOVE *Firefighters bravely risk their lives saving people from the burning building.*

office had been overturned at the gallery exit, which contributed to the pileup.

The fire blazed fiercely, lighting up the whole city. People were soon flocking to the scene by the thousands, looking for friends and family. The fire burned throughout the night, finally being extinguished about 5:30 a.m. Only the charred walls of the theater remained.

Crowds of mourning relatives claimed bodies from the temporary morgue set up in a hotel barn next door.

ABOVE *An hour after the beginning of the fire, the roof caved in. Only the charred walls of the theater remained after the fire was put out.*

The coroner declared that the unidentified bodies would be buried together on September 6th, although bodies were still being found on the 7th.

The Investigation

Captain Shaw, the special investigator, named twelve defects of construction, including the height of the ceilings over the auditorium, one staircase instead of two from the galleries, gas throughout the building being on one line, and no proper fireproof storage for scenery. The licensing authority of the one-year old building was heavily censured and the architects and contractor were also blamed. However, no charges were filed.

"It may be confidentially asserted that there can be very few worse, as the Theatre Royal."

—Captain Shaw, Special Commissioner of investigation

Seattle, Washington

DATE · JUNE 6, 1889
DAMAGE · $10,000,000
CAUSE · UNKNOWN

The fire began at 2:30 p.m., when some turpentine caught fire in the basement of a two-story wood framed factory building on Front and Madison Streets. It was a corner building of a row of attached frame buildings. Adjoining it was a wholesale liquor store, and as soon as the fire reached the barrels of liquor, they exploded with terrific force, scattering flaming firebrands for blocks. The fire quickly burned through

---

BELOW *The inferno raged out of control, and after two hours the water supply failed.*

blocks of warehouses and whipped by a northeast wind, jumped to Pryors Opera House, the largest on the coast. From there it leaped to the newspaper and publishing district, where the offices of the *Post Intelligence* were destroyed.

The fire department struggled to control the conflagration, but efforts were useless because, after two hours, the water supply failed. The reservoir was dry, and the firefighters did not have pumps to pump directly from the nearby bay. By 4:40 p.m. the flames roared into the

The City Before the Fire

Seattle was the largest town in the Washington Territory. It had a population of about 12,000 in 1885, and was the center of shipping for coal and lumber in the Territory. It was the terminus for the Columbia and Puget Sound Railroad, and the Puget Sound Shore Railroad.

ABOVE *The fire department struggled to control the conflagration, but after two hours the water for the hoses ran out, rendering all fire-fighting efforts useless.*

"It is Chicago repeated but on a smaller scale."

—*The New York Times*

The Damage

Every bank, hotel, office building, railroad station, warehouse and wharf was literally wiped out. Governor Miles C. Moore issued a proclamation saying, "The city of Seattle is in ashes. A hurricane of fire swept over the queenly city and she is in ruins. Thousands of her citizens are without food or shelter. Nothing can subdue the spirit of her people. She will rise again."

shopping district and quickly consumed stores and hotels, including the three-story Occidental Hotel, the finest hotel in the city. Block after block was burned, while panicked merchants, businessmen, shoppers and hotel guests fled from the fire. Several people died when the walls of Toklas & Singerman's building fell to the street.

In an hour firefighters arrived with equipment, via ship from Tacoma, and reinforced the exhausted Seattle firefighters. Buildings were dynamited in a futile attempt to create firebreaks, but blinding and suffocat-

ABOVE *The ruins of the Occidental Hotel, the grandest hotel in Seattle. During the fire, the owner, John Collins, tried to buy the buildings across the street, so they could be destroyed to form a firebreak. His efforts failed and the hotel was in ruins by nightfall.*

The Reaction

At a meeting following the fire, the city fathers decided to rebuild using brick and stone, hoping to fireproof the city against future fires. The streets were widened and the water and sewerage infrastructure was improved. There have been no more major fires since the rebuilding.

The Relief

Nearby cities immediately sent aid. Tacoma citizens sent large quantities of bread, provisions, tents and 100 slaughtered cattle. Portland sent several train-car loads of provisions, bedding, blankets and tents. Vancouver sent 70 tents. Monetary contributions from all over the country were funneled through former President Grover Cleveland and the New York banker Jay Gould.

ing clouds of smoke and flame roared onward through the business district, reaching the residential district by late afternoon.

By later that night the fire burned itself out, was contained and the next day Seattle was able to assess the damage. The burned district covered an area of thirty-one blocks, or sixty-four acres.

BELOW *Doctors James Eagleson and Clarence Smith set up their office in a canvas tent outside a house at Third Avenue and Columbia Street after the fire.*

Minneapolis, Minnesota

DATE · AUGUST 13, 1893
DAMAGE · 3 DEAD; $2,000,000
1,500 HOMELESS
CAUSE · UNKNOWN

ABOVE *A huge column of smoke rises above the south end of Nicollet Island.*

It was a little after 1:30 p.m. when a watchman saw a blaze on the river side of J. B. Clark & Co.'s box factory on the south end of Nicollet Island. Alarms were sounded, but by the time the fire department arrived, the building was engulfed in flames and lost. All the firefighters could do was try to protect the neighboring buildings. The flames, fanned by a furious south wind, swept through Lenhart's wagon wheel works and the Cedar Lake Ice Company. A spark carried by the wind fell on the lumberyard of Nelson, Tenney & Co. on Boom Island, nearly a half-mile away. The whole island was soon ablaze. Quickly jumping across the

Survivors

Approximately 100 people, principally laborers, were trapped near the mills by the flames. They saved themselves by jumping into the river, where they were forced to duck their heads underwater periodically to prevent the flames from scorching their faces, until they were rescued.

"Providence, and not the fire department, saved the greater portion of the manufacturing and residence portion of the East Side."

—*The New York Daily Tribune*

Mississippi River to the mainland, the fire burned through the lumberyard of E. W. Backus & Co. Neighboring mills fell, as the flames continued moving into town.

The mammoth brewery of the Minneapolis Brewing Company stood at Marshall Street and Thirteenth

Avenue, N.E. It seemed to the crowds of spectators on the bridges that it would be next, but the wind changed and the brewery, made of brick, acted as a firebreak. This prevented further spread of the fire northwards. Residents with garden hoses and buckets soaked their houses, but over 200 houses near the lumberyards burned.

St. Paul sent two engines in response to calls for help, but with the change in wind, the local firefighters were able to gain control over the fire. By 10 p.m. the fire was out. Hundreds of people slept in the open air the first night, but relief efforts began the next day.

The City Before the Fire

The site adjacent to the Falls of St. Anthony was first visited and named by explorer Louis Hennepin in 1680. Fort Snelling was established on the site in 1819, and Minneapolis was established on the western side of the river in 1847. It was incorporated as a city in 1867. The town of St. Anthony was annexed in 1872. Minnesota quickly became an early lumber center. Railroads and a growing flour-milling industry made the city an important hub by the 1890s.

BELOW *The Minneapolis Brewing Company stood at Marshall Street and Thirteenth Avenue, N.E. and acted as a firebreak. This prevented the fire from spreading further northwards.*

World's Columbian Exposition, Chicago

DATE · JANUARY 8, 1894
DAMAGE · 1 FIREFIGHTER DEAD; $2,000,000
CAUSE · UNKNOWN

The Exposition Before the Fire

The World's Columbian Exposition had been built in Chicago in 1892, in celebration of the four hundredth anniversary of Columbus's discovery of the New World. It was designed to inform visitors of the achievements of the United States in fine arts, industry, technology and agriculture. Designed by Daniel Burnham with Frederick Law Olmsted, it was built on the western shore of Lake Michigan. The buildings were little more than lath and plaster.

At 5:40 p.m. a park policeman saw a glow in a second story window of the casino, and before he had time to realize what was happening, flames burst through and crept toward the roof. He gave the alarm, but before a hose could be brought, 30 feet of the roof fell, as flames leaped into the sky. The fire was carried by the wind to the roofs of the Manufactures and Liberal Arts Buildings. The beautiful "White City" was ablaze with flames of red and gold.

Irreplaceable Loss

The Peristyle, with the Music Hall and Casino at either end, was the most imposing object seen by visitors approaching from ferries on Lake Michigan. In the center, forming the water gate to the grand Court of Honor, was the Columbus Portico, surmounted by "The Triumph of Columbus." A heroic statue of Columbus stood in a chariot drawn by four horses led by a woman. From each side of the Portico, 24 Corinthian columns, 14 feet high, formed a colonnade 234 feet long, to the Music Hall on one side and the Casino on the other. It was a sight to behold, and a shame to lose.

The Ruins

A second fire on July 3 destroyed what was left of the fair. The six large structures that formed the Court of Honor were destroyed. They were the Terminal Station, Administration, Manufactures, Electricity and Mining buildings, Machinery Hall and the Agricultural Building. A man, never caught, but suspected of arson, was seen running away from the grounds by the wrecking crew just before the fire broke out simultaneously in three buildings. This was basically the end of the White City, an idealized vision of America.

ABOVE *The tower on the Cold Storage Building in flames. The fires were extinguished during the night through the continuous work of the firemen.*

BELOW *Engine Company No. 60 with mascot Jim was one of the many fire-fighting teams on the blaze.*

"Sending rivers of regrets after it, we say…it was the fairest city that ever the sun shown on."

—General Lew Wallace

ABOVE *Smoke rises from the burning Agriculture Building, one of the six large structures that formed the Court of Honor at the exposition.*

Firefighters from Engine Companies No. 71 and 63 rushed to the roofs and had some success at containing the fire, as frantic exhibitors desperately tried to save their displays in the Manufactures Building. Firefighters were stationed inside to soak the floors and extinguish any dropping firebrands. However, portions of the plank roof caved in. The French exhibit, valued at $1,500,000 and containing Sevres porcelains, Gobelin tapestries, inlaid furniture and a collection of rare French books, was lost. The Japanese, German and Swiss exhibits were also totally destroyed.

By 9 p.m. the fire had spread to the imposing Peristyle. Engine Company Number 61 had raised an extension ladder to the roof, near the south end, and Firefighter William Mackey was the second man to climb up. He was carrying a line of hose to the roof, and in his haste to get to the fire, he missed a rung and fell to his death.

The fires were extinguished during the night, through the continuous work of the firefighters. Workers from the salvage company began to dismantle the buildings in the spring.

Hinckley, Minnesota

DATE · SEPTEMBER 1-2, 1894
DAMAGE · 413 DEAD;
  160,000 ACRES OF FOREST
CAUSE · LIGHTNING STRIKE

The summer had been very dry, and hundreds of acres of the pine forests in the Northern Plains were burning due to lightning strikes. The fires were being fought by farmers, but there was no organized fire-fighting force. Strong winds whipped the fire into a whirlwind, which

Fire Behavior

Fires have four rates of speed: Low, with a speed of less than 100 feet per hour; moderate, with a speed of 100–400 feet per hour; high, which moves from 400 to 800 feet per hour; and extreme, in which the fire moves 1,800 feet per hour, which was characteristic of the Hinckley fire.

tore across the prairies like the wildfires of today, such as those in Yellowstone or Los Alamos.

The town of Hinckley, Minnesota, seventy-five miles from St. Paul, with a population of 1,200, was wiped out of existence. Eighteen other towns including Mission Creek, Pokegama, Sandstone Junction, Cromwell, Miller and Bashaw were also swept away on September 1–2, 1894.

Newspapers reported pitiful stories from the stricken region of men, women and children running into small

Relief and Rebuilding

Monetary subscriptions were raised to aid the survivors. Orphans were placed with families or institutions. As the timber industry was no longer viable, the land was converted to farmland and given to anyone who had survived the destruction. Towns were rebuilt and today one can visit the Hinckley State Monument, which honors those who lost their lives in one of the greatest forest fires in national history.

BELOW *View of Hinckley's devastated main street the morning after the fire.*

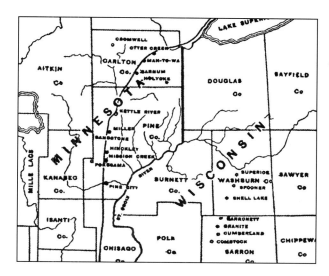

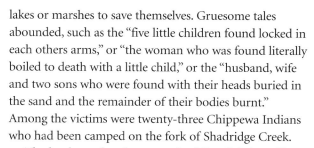

"Heaps of half-roasted bodies"

—*New York Daily Tribune*

lakes or marshes to save themselves. Gruesome tales abounded, such as the "five little children found locked in each others arms," or "the woman who was found literally boiled to death with a little child," or the "husband, wife and two sons who were found with their heads buried in the sand and the remainder of their bodies burnt." Among the victims were twenty-three Chippewa Indians who had been camped on the fork of Shadridge Creek.

The haphazard and unorganized fire-fighting efforts had little effect, but a driving rain finally came on the morning of September 3 and extinguished the flames. This only increased the misery, as people who escaped from the fires found themselves homeless and without food. Relief trains with food and drink were sent, but many refugees had lost everything.

ABOVE LEFT *Map of the burned district that included Hinckley, with a population of 1,200, and 18 other towns including Mission Creek, Pokegama and Sandstone Junction.*

ABOVE RIGHT *This engraving by Currier & Ives shows a train similar to the Duluth-St. Paul passenger train running before a wildfire.*

At the cemetery outside Hinckley, "ninety-six scorched, blackened, distorted bodies with bowels and brains protruding, hands clutched in a final agony" formed a heap through which desperate relatives searched for loved ones.

The Story of the Duluth-St. Paul Passenger Train

One of the most thrilling events associated with the fire was the experience of the Duluth-St. Paul passenger train, which was on its normal 2 p.m. express run. Dr. W. H. Crary reported to *The New York Times* that, "the smoke from the forest fires was so dense that lamps were lighted in the cars. The woods on either side were lashed by a fierce wind blowing at eighty miles an hour." When the train came within about a mile of Hinckley, the engineer found that he could proceed no further as the people were already fleeing for their lives from the town. The engineer was compelled to reverse the train and run backwards "leaving behind scores of others who sank to the ground exhausted and overcome by the terrible heat, never to rise again." The train had gone backwards only a short distance when "it was surrounded by a stream of flame which set fire to the engineer's clothing. He persisted and finally reached Skunk Lake where he shut off the steam." The passengers rushed out of the burning train and took refuge in the shallow water. "Many women had their clothes partially burned and torn from their bodies. One mother was found nursing her suckling child to prevent it from being suffocated." Miraculously, everybody survived.

Charity Bazaar, Paris

DATE · MAY 4, 1897
DAMAGE · 150 DEAD
CAUSE · MOVIE CAMERA EXPLOSION

The Bazaar Before the Fire

The flower of French aristocracy was assembled together for a charity benefit, in a temporary wooden building constructed to resemble a street of Old Paris. Each stall of the market, set up to sell goods for the charity event, was made of linen covered with papier maché and turpentine, making it water resistant, but highly flammable. Royal princesses, duchesses, countesses and other prominent leaders of society were in attendance. The aisles were thronged with visitors and purchasers.

The fire broke out soon after 4 p.m., above the stall of the Dowager Duchesse d'Uzes on the left side of the bazaar. The illuminating apparatus of the cinematograph (an early movie camera) exploded and set fire to the Turkish curtains and hangings. In a few moments the flames spread along the whole side of the bazaar, blazing through the turpentine and papier maché on the linen. A terrible panic and crush followed the sounding of a fire alarm. The bazaar had eight doors: three in front, one on the side, and four in the rear reserved for employees, all of which went unused because the royal guests did not know about them. There was a wild rush for the exits, and the weaker persons were trampled after having been knocked down in the stampede. The flammable nature of the building and its contents caused the fire to spread with great rapidity, and in a very short time the place was a mass of flame.

A policeman on duty said that from 1,500 to 1,800 people were in the building when the fire broke out. Shrieks of agony were heard as the fire swept behind the crowd struggling for the doors. In five minutes the heavily tarred roof fell in on the terrified crowd. The Papal Nuncio, who had blessed the opening a few minutes before, was unable to save the Duchesse d'Alençon, sister of the Empress of Austria. She was later identified by her jewelry.

Firefighters, unable to contain the fire, threw buckets of water on those who had managed to escape, but were still aflame. Naval cadets broke a few holes in the walls, but were able to save only a few people. Women whose skirts and petticoats had been burned or torn off were writhing on the ground. The dead were piled in heaps, especially near the exit, where the charred remains were five feet deep. Cabinet ministers, ambassadors, noblemen and members of society flocked to the scene in search of wives and daughters. By 8:30 p.m. the fire was out, and the search for bodies began.

An article in *Scientific American* urged that precautions be used in the future and that building regulations should be enforced.

"Why should a Bavarian Princess be burnt alive because she wished to raise a little money for the Catholic poor?" asked *The Spectator*, which gave its own response, "In the midst of life we are in death."

The Victims

The scene at the temporary morgue in the Palais de l'Industrie was horrific. Three long rows of bodies were covered with sheets. A cauldron of burning pitch was hung in the center from which people would light torches to view the bodies. Many bodies had horrible disfigurements, bones visible through fire-eaten flesh, some merely grinning skulls. In the center was a pile of coffins, and as a body was identified, it would be placed in one and sent home. Among the victims were a group of blind children who had been riding wooden horses when the fire began.

BELOW *Fire inspectors and police search the ruins for the remains of victims, digging for any clues that would help identify the badly charred bodies.*

The *New York Tribune* reported "the heads of the victims were burned to a cinder, even when other parts of the body were not much injured…as the roof fell in blazing masses on the heads."

Windsor Hotel, New York City

DATE · MARCH 17, 1899
DAMAGE · 33 DEAD, 52
INJURED; $1,000,000
CAUSE · CARELESSLY TOSSED
MATCH

The Building Before the Fire

The 7-story Windsor Hotel, built in 1873, was located on 5th Avenue between 46th and 47th Streets. The brick building cost $750,000 and could handle 600 guests. It was an older but comfortable hotel, in which there were permanent as well as transient guests, including President McKinley's brother Abner and his family.

Albert Johnson described the scene to the *New York Daily Tribune*: "A woman appeared at one of the windows on the sixth floor and started to climb out. A hoarse shout went up from the crowd for her to stay where she was but she disregarded it and sprang straight out. We saw her plunge down with her clothes pushed up by the descent. Then came the sickening crunch as she stuck the street and a dull groan ran all through the crowd…I saw a man climb out on the fire escape, fall off and strike every landing on the way down. He did not seem to have a whole bone left in his body."

The annual St. Patrick's Day Parade was just getting underway on March 17 when John Foy, a waiter, ran onto Fifth Avenue at 46th Street shouting "Fire, Fire!" In an instant the street was in an uproar.

The fire was started on the second floor by a carelessly tossed match. According to the *New York Tribune*, "An unknown guest of the hotel was standing in a bay window on the second floor which was hung with lace curtains. The guest lighted a cigar and tossed the match, still blazing, out the window towards the street. Just then the curtains blew back and caught fire from the passing match. The man who had thrown the match turned and ran out." Within minutes the fire spread to every floor, trapping on the upper floors people who, because of the noise from the parade, had not heard the cries of alarm from the hotel staff. Police Commissioners Abell, Hess and Sexton, riding in the parade in a carriage, were among the first on the scene. The hotel was surrounded on all sides by thousands of spectators who saw smoke and flames coming from the windows and panic-stricken guests leaning out the windows pleading for help. The first alarm was turned in at 3:14 p.m., and Engine Company No. 65 was the first to respond. By 3:28 a fifth alarm was sounded, and firefighters were raising scaling ladders and carrying helpless victims to safety.

By law, a rope long enough to reach the ground was located in every room, but many guests, lost in their panic, failed to use them. Chief Bonner said that he had never seen a more rapid fire. He knew at once that the building could not be saved.

Miss Helen Gould, living across the street in the Jay Gould mansion, threw open her doors to receive those in distress. "Bring anybody who needs help!" she cried out, "I'll turn the whole house into a hospital if necessary."

By sunset it was over. The walls and floors had collapsed, burying many victims in the ruins. The force of the fire flung pieces of masonry 100 feet down 46th

Rescues

There were many stories of heroic escapes, including one in which a sixteen-year old boy let himself down a rope with a baby under each arm. A father tied rope around the waist of one child and lowered him to the street, then hauled the rope back and did the same for another child, his wife and finally himself. Brave firefighters saved many lives, in one case passing an elderly woman from window to window to the fire-ladder. However, low water pressure hindered the efforts of the firefighters, and within a few hours the hotel was in ruins.

"The hotel was constructed under the old building laws of 1873. There was not a fire-proof thing in the place and absolutely nothing to check the spread of the flames all over the building once they gained a certain amount of headway."

——Fire Chief Bonner

ABOVE LEFT *The Windsor Hotel was constructed under the old building laws of 1873. By law, a rope long enough to reach the ground was located in every room, but in the mayhem many guests failed to use them.*

ABOVE RIGHT *The front wall of the hotel collapses onto Fifth Avenue as brave firefighters save many lives.*

The Reaction

The Windsor Hotel fire ranks as one of the worst hotel fires in New York City, and led to demands for more fireproofed buildings. Inspections of hotels and theaters worldwide followed the disaster.

Street. Over the next few days hundreds of searchers combed the ruins for remains of the dead and missing. At the East 51st Street police station there was a constant stream of visitors looking for lost loved ones. The dead included the wife and daughter of the hotel owner, Warren F. Leland.

North German Lloyd Lines, Hoboken, New Jersey

DATE · JUNE 30, 1900
DAMAGE · 326 DEAD;
 $10,000,000
CAUSE · BENZENE TANK
 EXPLOSION

The most destructive fire ever known in the Port of New York burned up the piers and three ocean-going liners at their berths in Hoboken.

The fire started when some cotton bales on one of the four large piers were set ablaze by the explosion of a benzene tank. The flames were spread by a high wind through all the piers with amazing swiftness. Discovered by the watchman at 4 p.m., within fifteen minutes not only four immense piers, but also three large ocean-going liners and several barges were aflame. According to the *New York Tribune*, "To many, it was unaccountable why the ships which were lying at the piers should have caught fire before it was possible even to cut them loose from the moorings. But all of the vessels had their portholes and doors open. Through these apertures, the flames swept with the fierceness of a

The Ship

The *Kaiser Wilhelm der Grosse* was then the fastest passenger liner operating between New York and Europe. She had won the North Atlantic Blue Riband in 1897 with a speed of 22.29 knots.

the ships, where men were imprisoned by the flames burning the upper parts of the vessels. They tried to reach down to the water and wet blankets to stave off the effects of the heat, but they could not get their bodies through the portholes.

Tugboats hauled the *Kaiser Wilhelm* into the river, and the crew managed to extinguish the fire onboard. The other three ships were blazing from stem to stern when their moorings burned and they drifted out into the river. The *Saale* drifted down to the Battery, at the southern tip of Manhattan, at about 6:30 p.m., and a New York police boat saved about thirty burned crew members.

There was great concern that the floating ships remain in the middle of the river and not drift to the New York shoreline, which was lined with wooden piers. Thousands of spectators lined the shores on both sides of the river. Fireboats from New York City and New Jersey sprayed water on the burning ships and the smoke rising to the sky changed colors and created a ghastly entertainment for viewers.

Funerals for the German crewmen, as well as the unidentified dead, were held at a North Bergen Cemetery. Private ceremonies for local longshoremen were held over the next few days.

ABOVE *The fire burned the moorings of the Saale, which drifted down to the Battery at about 6:30 p.m. A New York police boat saved about 30 crew members.*

forced draught. In addition, all the hatchways were open and through these, huge masses of burning cotton and flaming timbers fell into the holds and onto the lower decks." On piers 1 and 3 were several hundred casks of whiskey, which exploded, and added to the fury of the fire.

The fire trapped gangs of workmen and crews on the vessels. It was estimated that there were 500–600 longshoremen working on the piers, 400 men on the *Kaiser Wilhelm der Grosse*, 300 on the *Saale*, 250 on the *Bremen* and 300 on the *Main*.

The most dreadful scenes took place in the holds of

The Ruins

The charred hulks of the *Bremen* and the *Main* lay side by side after being towed back to the docks. The *Saale* rested on Liberty Island, under the gaze of the Statue of Liberty. A temporary morgue was set up and divers brought in the bodies of the drowned, while searchers onboard the three ships found many charred bodies.

The Reaction

An editorial in *Catholic World* deplored the construction of the piers and urged the federal and state governments to regulate fire-fighting on the piers, as well as revise naval architecture standards to increase fireproofing onboard ships. Fireproof interior walls using metal lath rather than wood became standard.

Atlantic City, New Jersey

DATE · APRIL 3, 1902
DAMAGE · $1,000,000
CAUSE · OVERHEATED BOILER

"Just what I have long expected."
—local resident quoted in the
New York Daily Tribune

The streets familiar from the game of Monopoly, such as Boardwalk, New York and Illinois Avenue, were in flames on April 3. Twelve hotels, along with many other businesses were destroyed in a fire, which swept the beachfront resort.

The fire was discovered shortly after 9 a.m. at Brady's Baths, which adjoined the Hotel Tarleton at Illinois Avenue, probably caused by an overheated boiler. The flames raged for nearly five hours, with such intensity that, at times, the whole city was threatened. All the burned buildings were wooden frame structures and directly adjoined the buildings on each side of them.

The fire, fanned by a strong southwest breeze, leaped from building to building with amazing rapidity. In fast

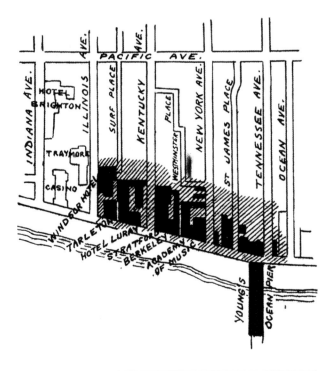

ABOVE *The fire destroyed seven blocks of prime real estate on the waterfront.*

succession the Luray, Stratford, Berkeley, Bryn Mawr, Tickney, Evard, New Holland, Rio Grande and Academy Hotels caught fire. The local firefighters were well organized, but they were unable to contain the fast moving fire, with the wood buildings providing plenty of fuel to burn. The firefighters wisely called for assistance, a move which was quite possibly responsible for the fire's ultimately ending with no loss of life.

Camden sent two fire engines, which were transported by special train via the Pennsylvania Railroad. Extra efforts were made by the firefighters to prevent the destruction of the Windsor Hotel, which was the last to catch fire. The wind had been favorable to the firefighters, but at 1:30 p.m. the wing of the Windsor nearest the blazing buildings began to burn, and was destroyed in a half hour. The rest of the hotel, thoroughly soaked by the firefighters' hoses, was saved, although it was badly damaged by smoke and water.

During the fire wild excitement prevailed among the guests of the hotels in the fire's path. Most guests had only a few minutes to pack some clothes and carry them to the beach, which appeared to be the safest location for fleeing families. As is common in these situations, thieves were busily robbing right and left until patrols of the National Guard were installed to guard the beach.

Just as the special train bringing firefighters and equipment from Philadelphia arrived, the fire jumped to Young's Pier near Tennessee Avenue, but was caught early, and with a strong force of firefighters, it was con-

ABOVE *View from the boardwalk looking west. People rummage through the ruins looking for belongings. The rollercoaster in the background survived the fire.*

tained to a loss of only half the pier. Nothing was left of the Boardwalk except the iron piers supporting it. By 3 p.m. the fire was out, thanks to the tremendous efforts of the firefighters.

The Rebuilding

Fortunately, evacuation efforts were successful and no lives were lost. Within a day, businesses were already clearing the charred ruins and rebuilding. Over the next few days, thousands of people came from New York and Philadelphia to witness the destruction and buy trinkets, which had been melted into freak shapes by the "Big Fire."

Iroquois Theater, Chicago

DATE · DECEMBER 30, 1903
DAMAGE · 588 DEAD
CAUSE · SPOTLIGHT IGNITING
A STAGE CURTAIN

"I extend to you, the people of Chicago, my deepest sympathy in this terrible catastrophe which has befallen you."
—President Theodore Roosevelt

The Building Before the Fire

The Iroquois Theater was billed as "modern and fireproof." It had just been built that year. Construction materials included "fireproof" ceramic blocks, electric lighting, not gas, and a fire resistant asbestos stage curtain.

The worst fire in American theater history occurred during a matinee performance of *M. Bluebeard*, starring the comedian Eddie Foy. The house was filled with an audience of about 2,000, mostly women and children. This was 250 more than the seating capacity, so the standing room area was filled.

Onstage eight people were singing, and all the lights in the auditorium and stage area were out. The scene was lit by a spotlight, which was at the proscenium, about 14 feet above the stage. A breeze blew a side curtain (not the fire resistant stage curtain) onto the spotlight. The spotlight was so hot that it ignited the curtain. The spotlight operator tried to extinguish it with his hands, but the flame ran up the

material and sparks began to fall on the stage. The
actors were alarmed, and ran to the back of the stage
to make their escape through the stage door to
Dearborn Street.

In an effort to contain the fire someone dropped the
asbestos curtain, but it got caught partway down on a
steel light reflector that had been left open. When the
stage door was opened, the draft fanned the flames
towards the ventilator at the rear of the balconies. All
this happened before the audience was even aware of
the danger.

The stage was a roaring furnace and the auditorium was still in total blackness, except for the low
orange light from the fire. The audience stampeded
toward the front doors, ignoring many other exits.
The whole theater was soon filled with hot suffocating
smoke, the upholstery of the seats caught fire and the
few exits, which the audience attempted to use, were
blocked by fallen people. Most of the victims were in
the balconies.

ABOVE *Panorama of Iroquois Theater after the
fire.*

The Victims

Many of the victims were children, and the scene at the makeshift morgue
was pitiful. *The New York Times* describes bodies stretched in long lines on
the floor as "tearful men and women would lift the sheets from the charred
and bruised faces searching in the clothing of the dead for something by
which their loved ones could be recognized.... One of the most peculiar of
the identifications was that of the headless body of Boyer Alexander, 8 years
old, who was identified by his father from the watch he had been given for
his birthday."

BELOW LEFT *Drawings from the front page of an extra edition of a Chicago newspaper depict attempts to escape from the balcony. Police and fire officials (at right) search the remains for victims.*

BELOW RIGHT *Fireproof building materials made from metal were mandated for public buildings following the Iroquois Theater fire. The "Kno-Burn," "Py-ro-no" and "The City Unburnable" manuals show fireproof metal lath and asbestos construction material developed at the time.*

Fire Resistant Asbestos

Asbestos is a grayish mineral, which occurs in long, thread-like fibers. It will not burn and was used at this time in fireproof curtains, roofing and insulation. Now, we know that the fibers are hazardous when breathed, and can cause a lung disease known as asbestosis.

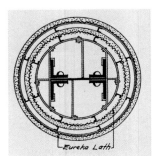
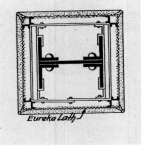
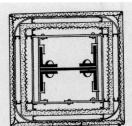

Eureka Lath   *Eureka Lath*

Eureka Lath Plaster   *Eureka Lath*

Fig. 43

The Investigation and Reaction

Mayor Carter H. Harrison proclaimed that there would be no New Year's Eve celebration in the city and Chicago silently mourned its dead. Nineteen Chicago theaters were ordered closed on January 1, 1904, pending inspections. An investigation following the fire showed that there were no fire hoses or sprinklers in the theater and the firefighter stationed in the theater (a common practice at the time) had deserted his post.

The Iroquois Theater fire brought nationwide reforms in theater construction codes, more fire walls, more and better exits, unobstructed alleyways and fireproof scenery. Safeguards such as sprinkler systems and fire hoses were standardized.

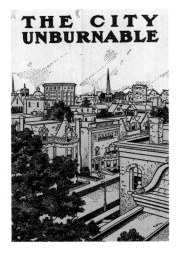

PYRONO Door and Trim, showing portion with veneer cut away and magnified, disclosing Asbestos Sheathing indented into wood core.

THE CITY UNBURNABLE

Baltimore, Maryland

DATE · **FEBRUARY 7-8, 1904**
DAMAGE · **$100,000,000;**
1600 BUILDINGS,
86 BLOCKS, 1 DEAD
CAUSE · **UNKNOWN**

Baltimore's "Great Fire," second only to the Chicago fire in its magnitude and the financial losses involved, raged for thirty hours in the heart of the banking, wholesale and warehouse district.

BELOW *Flames shoot out of the upper windows of a Baltimore building. More than 1,200 firemen fought the flames to no avail, until a shift in the wind helped them contain the fire at the waterfront.*

Fire Breaks

In a large urban fire, firefighters look for natural barriers where the fire will have less fuel and slow down to the point where fire hoses can extinguish the flames. They try the best they can to guide the fire to these areas, control the sides of the fire by dowsing unburned buildings in water, and set up for a final massive dowsing at the fire break. Battling the fire head on, when several buildings are aflame and throwing tremendous heat, is largely futile. Rivers, waterfronts, large brick or stone structures and wide roads make good fire breaks. Sometimes firefighters create a fire break by blowing up buildings in the fire's path, depriving it of fuel.

The fire started on Sunday morning at 10:45 in the dry goods store of John E. Hurst & Co., at the intersection of German and Liberty Streets and Hopkins Place. Churchgoers heard a large explosion of unknown origin, and rushed outside to a rain of cinders, which quickly ignited building after building.

A gale of 30 miles per hour was blowing in a northeasterly direction, sending the flames through the crowded streets. Frantic business owners sent hundreds of wagons into the area to try to save their books, files and safes, which only served to clog the streets and delay firefighters from reaching the flames at the beginning of the fire—the most critical time. Within half an hour, a dozen big warehouses were burning fiercely. By noon, thirty buildings were blazing, despite being

BELOW *The National Guard stands in front of a building that is covered in ice from the water sprayed by firemen. The burned district covered almost 140 acres and destroyed 86 blocks.*

soaked with water by firefighters as a preventive effort.

The valiant firefighters were helpless to stop the spread of the flames, and called for reinforcements. Washington sent six engines at 1 p.m., Philadelphia sent four engines and seven engines came from New York. More than 1,200 firefighters fought the flames, but by 7 p.m. the situation was hopeless.

Chief Fire Engineer George W. Horton decided that the only thing left to do was dynamite buildings at key points to serve as firebreaks. Buildings on South Charles Street were blown up, but this only delayed the flames and the fire continued to spread. With the destruction of the central telegraph office, Baltimore was cut off from communication with the rest of the country. President Theodore Roosevelt ordered members of the Washington Police Bureau to assist in any way possible.

At 8 p.m., with the fire burning for more than nine hours, the fire was still spreading and there was a tremendous explosion of about 150 barrels of whiskey stored on the upper floors of a warehouse at 24

ABOVE *The fire's destruction was immense, as seen in this photograph taken a few weeks after the fire. Over 1,525 buildings burned including banks, hotels, office buildings and warehouses.*

Hanover Street. Two fire wagons caught fire and a fire engine was buried by a falling wall. During the night, thousands of people watched the spectacle from hillsides surrounding the city. *The New York Times* reported, "The pyrotechnic display has been magnificent and imposing beyond the power of the painter to depict. Vast columns of seething flame are shooting skyward at varying points of the compass and the firmament is one vast prismatic ocean of golden and silver hued sparks."

On the morning of the 8th, the wind shifted to the southeast; firefighters were at last able to take advantage of the waterfront and the Jones Falls stream as natural obstacles to the fire. The fire was finally controlled later that afternoon.

"This is one of the greatest calamities this country has ever known."
—Senator Chauncey M. Depew

The Ruins and Reaction

The burned district covered almost 140 acres and destroyed 86 blocks. Over 1,525 buildings burned, turning out 2,400 business, including banks, hotels, office buildings and warehouses. The loss of life was limited to only one person, but the damage was estimated at $100,000,000.

Almost before the fire was out, the people of Baltimore were talking about improvements. The elimination of above-ground electric wires, street widening, additional parks, better docks and wharves and an improved sewerage infrastructure. The fire brought about stricter building codes and the use of fireproof materials, such as hollow ceramic tile blocks or concrete block and steel.

General Slocum Steamship

DATE · JUNE 15, 1904
DAMAGE · 1,021 DEAD
CAUSE · UNKNOWN

The worst loss of life in a single fire in New York City and one of the worst disasters in maritime history occurred during a summertime Sunday school picnic excursion.

The congregation of St. Mark's German Lutheran Church on East 6th Street was in a festive mood as they boarded the excursion vessel for their annual picnic at Locust Grove on the north shore of Long Island. They were mostly women and children under the kindly care of their pastor, the Reverend

The Ship Before the Fire

The wooden vessel had three open decks and could accommodate 2,500 passengers. She had been built in 1891 and was named for a Civil War general. Coal-burning boilers powered her two steam-driven, side paddle wheels.

Hundreds of dead were piled on the shore of the island. A witness reporting to the *New York Tribune* described the scene: "It looked like a battlefield. Heaps of dead were piled here and there, sometimes as many as fifty lying in a heap. I treated as many injured as I could and sent them to the mainland in charge of nurses."

BELOW *The Steamship* General Slocum, *before one of the worst disasters in maritime history.*

RIGHT *Hundreds of dead were piled on the shore of the island. Bodies were being recovered at the rate of one per minute.*

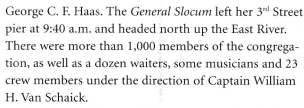

The Reaction

As a direct result of the tragedy, regular inspections of steamboat safety equipment, such as lifejackets and fire hoses, were mandated. Wooden excursion ships began to be replaced with steel vessels.

George C. F. Haas. The *General Slocum* left her 3rd Street pier at 9:40 a.m. and headed north up the East River. There were more than 1,000 members of the congregation, as well as a dozen waiters, some musicians and 23 crew members under the direction of Captain William H. Van Schaick.

The fire was discovered at 9:56 a.m., under the forward boilers on the port side, and after several minutes attempting to put it out using rotten canvas hoses, the crew notified Captain Van Schaick of the problem. The boat had already entered the treacherous, fast-moving waters of Hell's Gate at the northern end of Manhattan where the East River joins the Long Island Sound and the Harlem River. Captain Van Schaick decided to head the burning boat at full speed for North Brother Island, another mile and a half north. The speed of the boat fanned the flames, sending them roaring along the lower deck. Clouds of smoke rose to the upper decks, obscuring them. Hundreds of terrified women and children rushed to the stern of the boat, hoping to avoid the flames. The crush forced many passengers against the railing, which broke, tossing people into the swirling water of the dangerous confluence. The captain never gave the call to abandon ship, possibly because of the treacherous waters.

The canvas life jackets had not been replaced since

"When the fire broke out, the deck-hands knocked down everybody trying to save their skins."

—George Heinz, survivor

the boat was built 13 years earlier and they were rotten and useless. As they jumped overboard, women and children fell on one another, drowning before help could reach them. Several nearby tugboats raced to assist and plucked people out of the water, but could not get close to the burning ship, due to the heat of the fire.

Less than fifteen minutes after the discovery of the fire, the *General Slocum* went aground on some rocks nearly one hundred feet from the shore of North Brother Island. As she struck the rocks, the upper hurricane deck suddenly caved in, plunging passengers into either the flaming hold or the water. The water around the vessel was filled with drowning people. Three men from shore saved sixty of them, and nurses from the contagious diseases hospital on the island waded into the water with ladders to rescue victims, and then struggled to resuscitate them. Police and firefighters working from the tugs saved many of the congregation members.

Mayor George B. McClellan ordered an immediate investigation, and the surviving crew members of the *General Slocum* were arrested. Captain Van Schaick was convicted of neglect of duty, and sentenced to ten years in prison. He served three-and-a-half years in Sing Sing Prison, until pardoned by President Taft for good behavior.

The Organization of the Slocum Survivors was formed to commemorate the event and met yearly at the cemetery. The fiftieth anniversary service was held on June 13, 1954, and the white-haired men and women, who were flaxen-haired boys and girls at the time of the accident, bowed their heads in prayer.

San Francisco, California

DATES: APRIL 18, 1906
"GREAT FIRE"
DAMAGE · 28,188 BUILDINGS
DESTROYED; 497
CITY BLOCKS; 500
PEOPLE DEAD;
$1,000,000,000
CAUSE · EARTHQUAKE

In the gray dawn of the morning of April 18, at 5:15 a.m., a fierce earthquake rocked the city of San Francisco. The "City by the Bay" experienced one shock, which lasted two minutes. People became

ABOVE *Panoramic view near Turk and Market Streets.*

OPPOSITE PAGE, BELOW *The city in flames. The dense smoke that rose from the business district spread out over the town and could be seen for miles.*

panic-stricken and fled to the streets where many of them were killed by showers of bricks, cornices and walls.

Fires, erupting from broken gas mains, broke out simultaneously throughout the downtown area. The fire department promptly responded to the first calls, but the quake had burst the city's water mains rendering them useless. The fire swept down the streets so rapidly that it was impossible to save anything in its way. The commercial buildings and banks north of Market Street were burning and the flames swept down Mission Street and quickly consumed everything in their path, including the Grand Opera House. The Palace Hotel, in which Enrico Caruso was a guest, burned to the ground. Great sheets of flame rose high in the sky or rushed down narrow streets. The dense smoke that rose from the business district spread out over the town and could be seen for miles. The Post Office, the Grand Hotel, and City Hall were all destroyed by the fire.

The Richter Scale

The Richter Scale was developed in 1935 by Charles F. Richter, of the California Institute of Technology. According to the United States Geological Survey, the magnitude of an earthquake is determined from the logarithm of the amplitude of seismic waves. Since 1900, the largest earthquake recorded was in Chile on May 22, 1960, with a magnitude of 9.5. An earthquake in Alaska on March 28, 1964 was the second biggest in the world, and the largest in the United States, measuring 9.2 on the Richter scale. Examining older records, experts have placed the 1906 San Francisco earthquake at 7.7 magnitude.

Restoring Order

Relief stations were set up in Golden Gate Park and police and military kept strict control. The supplies in all grocery stores were seized, and drinking water was brought in from nearby Lake Merced. Thousands of families slept in the open air. When darkness fell over the desolate city, at the end of the fourth day, the exhausted survivors knew that the task of rebuilding San Francisco would begin the next day.

Without water, soldiers were ordered to blow up buildings to create fire breaks, but by the end of the first day, an area eight square miles had been burned, which created too large an area to contain by dynamiting buildings, and it was apparent that the entire city was doomed.

Thousands of people had fled to the hills, where they watched the spectacular flames consume the city. Chinatown and Japantown went up like tinder and a constant stream of refugees made their way from one area of town to another, fleeing the flames. The fire advanced through Union Square and moved up to Nob Hill, where it consumed the opulent mansions of the wealthy.

On the morning of the second day, General Frederick Funston and Mayor Eugene Schmitz imposed martial law, with orders to shoot to kill any looters. Refugees fled to the Presidio military base. By the third day the fire was raging over fifty acres of the waterfront, and the warehouses were soon in flames. The order was given to make a last stand at Van Ness

Avenue and, to ensure that the fire would not jump across, the magnificent mansions on the eastern side of the street were dynamited. At last, this seemed to work, and by the end of the third day, Mayor Schmitz issued a proclamation declaring that the fire was contained. However, he was premature, because a stiff wind blew the flames toward the Union ferry terminal, the last means of leaving the city. Hundreds of U.S. sailors joined the fight, and fire tugs threw immense streams of water from the bay onto burning warehouses. Fort Mason was saved by strenuous efforts.

During the week succeeding the fire, soldiers and police forced every man to help clear the ruins. Over 10,000 tents were set up and relief lines provided rations for 349,000 refugees.

The response of the nation to the pleas of the stricken city was immediate. People at all levels of society donated money and supplies. Congress appropriated millions of dollars, and President Theodore Roosevelt coordinated efforts through the American Red Cross.

"After four days and three nights that have no parallel outside of Dante's *Inferno*, the city of San Francisco … was a mass of glowing embers fast resolving into heaps and winrows of gray ashes emblematic of devastation and death."

—Opening line of the *Complete Story of the San Francisco Horror*, published in 1906

ABOVE LEFT *"The Stricken City," sheet music from 1906. The response of the nation to the call for aid was immediate.*

BELOW *Panorama of the Western Addition and the Mission, showing the ruins of City Hall and St. Ignatius College, with Twin Peaks and Strawberry Hill in the distance.*

Rhoads Theater, Boyertown, Pennsylvania

DATE · JANUARY 13, 1908
DAMAGE · 170 DEAD
CAUSE · OIL LAMPS IGNITING A STAGE CURTAIN

One fifteenth of the population of this small town was killed in one night at the Rhoads Theater. Fifty members of the St. John's Lutheran Church had just presented a five-act play with tableaus, entitled *Mary, Queen of Scots*. The audience of 500 friends and family were going to see a new innovation, a motion picture, using a new machine, the stereopticon. The operator, Henry W. Fisher, turned the valve and a rush of hissing gas escaped. It was so loud that it scared the audience. The actors attempted to quiet the crowd by telling them

BELOW *The smoldering interior of the Rhoads Opera House. It was a three-story frame building, with a brick front. The auditorium occupied the second and third floor, while the first floor had businesses.*

The flames spread with startling rapidity and a terrible panic ensued. A survivor was quoted in *The New York Times*, "It was as sudden as it was unexpected...the stampede, the panic-stricken people wedged in the stairway, the cries of those in the rear, the dreadful smoke, the women and children overcome...and then scores engulfed by the seething fire."

LEFT *The proportion of victims was about nine females to one male.*

> "It all took place in less than two minutes...the terrible cries of the people I will never forget even to my dying day."
> —survivor

everything was okay, but inadvertently kicked over the oil lamps that had been in use as footlights. The flimsy curtain caught fire and a sheet of flame ran up the scenery to the roof and engulfed the stereopticon.

Panicked men were reported to have jumped from their seats and, seizing chairs, beat down women and children in their rush for the exits. A bolted exit door gave way under the force of the pushing crowd, and those behind trampled those in the front. There were only a couple of windows and some people managed to save themselves by jumping. However, the rush to get to the windows caused a pile of writhing bodies, which was nearly impossible to climb.

The proportion of victims was about nine females to one male. In almost every case, the upper portions of the bodies were burned away, the lower portions below the waist were intact, showing that the members of the audience were wedged in when the flames swept over them.

The single Boyertown fire engine had responded in five minutes, but because of an accident to the pumping mechanism while going down a hill on the way to the fire, firefighters could not throw any water on the building. Calls were made to Pottstown and Reading, but within fifteen minutes not a vestige of the building was left, except the blackened walls.

By 6 a.m. the fire was under control, and the first bodies were brought out at 8:30 a.m.

The Theater Before the Fire

The Rhoads Theater was a three-story frame building with a brick front. The auditorium occupied the second and third floor, while the first floor had businesses. At the rear and part way around the sides was the balcony. The motion picture machine was directly in front of the only stairway, which cut off any means of escape.

Lake View School, Collinwood, Ohio

DATE · MARCH 4, 1908
DAMAGE · 176 DEAD
CAUSE · FURNACE

"They were crushed so tightly together that no human strength could clear a passageway. Dozens of them died within a foot of absolute safety."
—Miss Anna Moran, School Principal

The worst school fire in the United States occurred at the Lake View School, in a suburb of Cleveland, on March 4, 1908.

At 9:40 a.m., when the flames were discovered shooting from the furnace, the teachers marshaled the students into lines for a fire drill. Unfortunately, they had always practiced leaving by the front door, and when the children reached the foot of the stairs, flames surrounded them.

The front and rear exits had two sets of doors, with a vestibule between them. They were open, and a few dozen children managed to escape through the front doors before the flames completely blocked the them, burning up through the floor. Children coming downstairs from the upper floors were descending right into the flames. The children at the foot tried to fight their way back up the stairs, but were shoved mercilessly back down into the flames below or were crushed under the heels of their panic-stricken playmates. Bodies were now piling up in the flames at the foot of the stairs.

The rear doors were open, but children had fallen down the stairs, blocking the passageway. A huge heap

The Heroes

Miss Laura Bodey, fifth grade teacher, saved almost fifty of her class from the third floor, by bravely taking them through smoke and flames down the fire escape. Some frightened children in the rear fled back upstairs and perished in the flames. One witness told *The New York Times*, "I can see the wee things holding out their tiny arms and crying to me to help them...their voices are ringing in my ears yet." Two teachers died in the flames trying to help their pupils to safety.

of still-living children, nine-bodies deep and almost reaching the ceiling, was trapped at the doorway. The crowd, which had gathered outside, saw a mass of white faces and struggling half-burnt bodies.

It took about twenty minutes for firefighters to arrive, but then they did not have a ladder that reached the third floor, and the spray from the fire hoses could not reach the upper floors where the children were trapped. Within a half hour after the alarm had been sounded, hundreds of parents arrived on the scene and fought frantically with police and firefighters to enter the burning building, further hampering firefighting efforts.

The bodies were taken to the makeshift morgue, located at the Lake Shore Railroad Storehouse, and arranged in rows of ten, but Deputy Coroner Harry McNeil declared that the faces of nearly all the children were so badly burned that it would be impossible to identify many of them. "I have portions of bodies and dozens of hands and feet which have been torn off," he said. A 9-year-old was known only by a fragment of his sweater. An 8-year-old was identified by a little pink-bordered handkerchief, in which he had wrapped a new marble. Another was identified by his suspender buckle. By 10 p.m., sixty bodies had been identified. Dawn on the 5th saw mothers and fathers digging in the ruins for their missing children. Burials of the identified dead began on the 6th. The rest of the unidentified dead were laid side by side at the Collinwood Cemetery.

The School Before the Fire

The school, built in 1901, was a wooden three-story structure, with a furnace in the basement, under the front stairway. The number of students was more than normal, and the smaller children had been placed in the upper floors. There was only one fire escape, in the rear of the building, and there was a second stairway, also in the rear.

HISTORICAL ACCOUNT

Historian Edward (Sonny) Kern recounts, "By 11:00 a.m. with all hope gone, the Cleveland and Collinwood Fire Departments remained principally to pour water into the still smoldering ruins. Once the fire was completely extinguished by the firefighters, the supreme horror of digging through the debris in the basement for the bodies of the children had to be done."

Triangle Shirtwaist Factory, New York City

DATE · MARCH 25, 1911
DAMAGE · 146 DEAD
CAUSE · UNKNOWN

At 4:40 p.m. a fire of unknown origin broke out on the 8th floor of the Asch building on one of three floors occupied by the Triangle Shirtwaist factory. An attempt by workers to extinguish it with buckets of water failed

Thousands of spectators who had come from Broadway or Washington Square were screaming with horror at what they saw. According to eyewitness accounts in *The New York Times*, "Five girls stood together at a window on Greene Street then leaped together, clinging to each other with fire streaming from their hair and dresses. They struck a glass sidewalk cover and crashed through to the basement... A girl on the eighth floor leaped for a firefighter's ladder, which reached only to the sixth floor but missed it and fell to the street."

The Factory Before the Fire

The infamous Triangle Shirtwaist factory was located on the top three floors of a ten-story building on the corner of Greene Street and Washington Place in Greenwich Village, adjoining a New York University building. The building itself was "fireproof," but each floor was filled with sewing machines packed so closely together that it was nearly impossible to get between them. Hanging from rope lines, stretched across the space above the heads of the immigrant workers who were mostly girls 16 to 23 years old, were hundreds of shirtwaists on which they were laboring. Piled on the floors were heaps of remnants and cuttings created by the workers. There were only two freight elevators and one internal stairway. There were no exterior fire escapes.

RIGHT *Firefighters working to douse the flames at the Triangle Shirtwaist Co. on Washington Place. The firemen had trouble bringing their trucks into position because the bodies of people who had jumped out of windows were strewn about the sidewalks and streets.*

OPPOSITE *Police examine one of the doors at the factory which the owners had illegally locked to prevent theft. Had the door been open, many of the victims could have escaped down the stairway.*

Rescues

Valiant and quick thinking students from the adjoining NYU Law School rushed to the roof of their building and lowered a ladder to the roof of the factory building. They were able to save nearly one hundred women who had escaped to the roof.

because too many piles of cloth were already aflame, some as high as the workers' waists. The flames spread with deadly rapidity, fueled by the flimsy fabric used in the factory. Four alarms were rung within fifteen minutes. Girls screaming in Italian, Russian, Hungarian and Yiddish tried to use the stairway, but found it blocked by a locked metal gate that the owners had illegally installed to prevent theft.

Flames were pouring out of the building, and panicked girls began to leap out of the windows even before the first fire engine arrived. The bodies of the

The Reaction

The Triangle Shirtwaist fire led to better working conditions for women, including passage of a 54-hour work week, higher wages, enforced fire codes and the right to unionize. The fire contributed to the growth of the International Ladies' Garment Workers' Union (ILGWU), furthering the relationship between labor and the clothing industry. It also led to the idea of adding sprinklers to buildings for the safety of workers.

LEFT *The Triangle Shirt factory fire horrified the nation, sparking the movement for better working conditions for women.*

jumpers blocked the firefighter's efforts to bring the trucks into position.

Many jumped into fire-nets or blankets, which proved useless because so many leaped at once that the nets were pulled from firefighters' hands.

It was all over in a half hour. Most of the victims were suffocated or burned to death within the building. Piles of dead were found near the elevators and at the locked doors. Fifty burned and mangled bodies were taken from the ninth floor alone. Bodies were lined up in lidless boxes on the street, awaiting transfer to the morgue.

"It's human, that's all you can tell."
                        —policeman gazing at a victim

The Investigation

The owners, Max Blanck and Isaac Harris, had escaped unharmed. They were later charged with manslaughter, due to the unsafe working conditions. However, they beat the charge with the help of a clever lawyer, and collected insurance money. After years of court hearings they ended up paying off lawsuits at just $75 per body. They continued to flaunt fire laws in their new factory, where inspectors again found illegally locked doors.

Dreamland, Coney Island, New York

DATE · MAY 27, 1911
DAMAGE · $5,000,000; ONE
      FIREFIGHTER DEAD
CAUSE · ELECTRIC LIGHTS
      EXPLODING

Coney Island had been open for the summer season for just one week when a fire started sometime after midnight at Dreamland, one of three popular amusement parks in the area. The electric lights over Hell Gate, a boat ride through dimly lit caverns, exploded and set a knocked over bucket of tar on fire. The first alarm was called in at 2 a.m. and the firefighters arrived quickly. However, the newly installed water pumping station

---

RIGHT *People on the boardwalk at Dreamland, Coney Island before the fire. It was dominated by the monumental Dreamland Tower, designed by architects Kirby, Petit & Green, which was destroyed in the fire.*

BELOW *Dreamland at night. Coney Island featured the Dreamland Tower, lit with the new electric light.*

which had boasted water pressure of 125 pounds produced only 20 pounds, comparable to the pressure of a garden hose. It was thought that the connection between the new system and the water of the Atlantic had been faulty. Meanwhile, the fire grew out of control.

Finally, at 4 a.m. the "two nines" alarm was sounded for the first time in the history of Brooklyn. It brought firefighters from Queens and Manhattan.

Before the Fire

Coney Island first became a popular resort with the opening of the Hotel Brighton in 1876. The opening of George Tilyou's 15–acre Steeplechase Park in 1897 attracted a middle-class crowd looking for enjoyment. Luna Park, opened in 1903 by Frederick Thompson, featured fantastical towers lit with the new electric light. Dreamland was opened in 1903 by Senator J. R. Reynolds, a real estate investor.

The "Incubator Babies," a specialty show featuring five premature babies each weighing as little as 14 ounces, was next to where the fire began. The night watchman alertly roused the nurses. His quick thinking allowed the nurses to save the babies. The wild animals at Ferrari's Animal Show, which had delighted thousands of children, were not as fortunate. "Little Hip," the elephant, trumpeted her fear until waves of flame took her. The howls of the animals were pitiful to hear, as monkeys, lionesses and ponies perished in the

ABOVE *The concessionaires immediately came in and began to clear a space to set up tents for exhibits in the ashes.*

BELOW *New York's first gasoline-driven pumper, acquired in 1911. This pumper was not very successful and broke down often. The propulsion system was that of a standard automobile of the time.*

flames. "Sultan," one of the largest lions, escaped when his cage crumbled, and with flaming mane he burst through the "Creation" gateway and ran down Surf Avenue seeking safety. Thousands of spectators who had gathered to watch the fire shouted "the lion's loose" and scattered. Sultan climbed to the pinnacle of the "Rocky Road to Dublin" exhibit, which resembled a mountain, where he was killed by an axe thrown by a fireman. His body plunged into the crowd below, which tore him apart for souvenir teeth and claws. Captain

Fire Alarms

Fire alarms at this time were keyed to a physical location of the box from which they were rung. Thus an alarm of 3–3–2–174 would tell a nearby fire company that 3–3 was a 3–alarm fire and 2–174 was the Dreamland box. The 2-nines alarm of 9–9–2–174 meant it was the largest fire ever in Brooklyn. The 2–nines alarm called out 33 fire companies and over 250 firefighters. Today, this type of alarm system is no longer in use. Fire alarms are wired directly to a central fire control room that directs firefighters to fires.

"Our pumps worked well, but they were overmatched. It was a hopeless job. The blame can only rest directly upon the executives of the Fire and Police Departments for not using the water within the plant's capacity of twelve lines of hose."

—Water-pumphouse Chief Engineer Baroni

An Unlikely Survivor

Two weeks after the fire, a 350-pound lioness named Atlanta was found alive, living under the Luna Park Boardwalk. She had survived by feasting on pigeons and chickens that were kept to provide eggs for employees of Luna Park. Captain Ferrari demanded her return, but the owner of Luna Park decided to exhibit her to earn back the costs of the chickens.

Ferrari, the lion's owner, had to fight the rabid relic hunters to save the head and body.

By 5 a.m. the fire burned itself out, and by 7 a.m. the firefighters were heading home. It had burned its way to the end of the island on the east, but had been contained near the Giant Steel Roller Coaster on the west and the ocean on the south. The buildings on the south side of Surf Avenue were completely destroyed, including Balmer's Baths and all the souvenir shacks on the New Iron Pier. The concessionaires immediately came in and began to clear a space to set up tents for exhibits in the ashes.

A committee of New York City Aldermen recommended the purchase of the burned-out site, and in July the Board of Estimate voted to acquire the property and convert it to a beach and park.

Coney Island became even more popular with the extension of the New York subway in 1920. In 1923 the Boardwalk opened, and 175 businesses were razed to widen the streets. In 1927, the Cyclone roller coaster opened, and the resort reached the height of its popularity. But fires continued to plague Coney Island, and with the burning of Luna Park in 1944 and the burning of the Steeplechase Pier in 1957, the downward spiral began, which culminated with the closing of Steeplechase Park in 1964. The wooden Boardwalk remains, as well as a few small sideshows, also built of wood, so the risk of fire at Coney Island remains.

Equitable Life Building, New York

DATE · JANUARY 9, 1912
DAMAGE · 6 DEAD, $2,000,000
CAUSE · UNKNOWN

The fire was discovered about 5:30 a.m., in the basement kitchen area of the Café Savarin, in a rubbish heap. Employees struggled to put it out for a half hour before sounding an alarm. The nearby elevator shaft acted as chimney and drew the flames in, and a few minutes later the fire had reached the top floor. The outside temperature was below 20 degrees, with a sixty-mile an hour gale blowing when Chief Kenlon and his firefighters arrived. Ice seemed to form in midair as the firefighters tried to hose down the flames and unburned parts of the building. The ice clogged the machinery and froze the hoses to the ground. It settled

The Building Before the Fire

Henry Baldwin Hyde founded the Equitable Life Assurance Society of the United States in 1859, and was its president until 1899. The company had offices in Paris, Vienna, Berlin, Madrid, Melbourne and Sydney. Gilman and Kendall constructed the New York headquarters building in 1870, at 120 Broadway, with the architect George B. Post, in a mansard-roofed Empire Style.

ABOVE LEFT *The grand interior of the Equitable Life Assurance offices before the fire. It was built in 1870.*

ABOVE RIGHT *The Equitable fire ruins, Broadway and Cedar Street. In some places, two feet of ice covered every surface of the charred building.*

OPPPOSITE PAGE, ABOVE *The extremely low temperatures clogged fire engines with ice and froze the hoses to the ground. Despite the icicles, this fire engine continued working through the blaze.*

OPPPOSITE PAGE, BELOW *The temperature dropped below 20 degrees with a sixty-mile-an-hour gale blowing when firefighters arrived.*

"It has been a great battle made all the harder because of narrow streets, weather that approached the zero kind, and the fact that it was in an old building."
—Fire Chief John Kenlon

Police had to be called to fight back the crowds on Nassau Street, which were hampering the fire-fighting efforts. One of the strangest spectacles that kept the growing crowds fascinated was the sight of two despairing hands thrust through the bars of the basement vaults. They reached out in vain appealing for help. *The New York Times* wrote: "It was all of the man that could be seen… and as the afternoon drew to a close, the ice mound on the pavement grew layer by layer, till even the hands were hidden from view." *The New York Times* also reported "ice converted the ruined building into a fantastic palace with the rainbows arching at every turn as the sunlight filtered through the spray and the smoke." In some places two feet of ice covered every surface.

on the men themselves, and their boots periodically had to be chopped out of ice heaps so they could keep working.

Firefighters sprayed water from the roofs of nearby buildings and managed to contain the flames by evening. The Lawyers Club and the Law Library, containing 22,000 volumes, were destroyed and court cases were delayed due to lost papers. Other tenants found offices nearby, and Equitable Life announced plans to rebuild. Billions in securities and bonds housed in vaults in the basement were safely recovered within a couple of days.

The bodies of the dead, including Battalion Chief William Walsh, were recovered from the masses of ice and twisted iron a few days later. The funeral of Chief Walsh was attended by 2,000 mourners, who crammed the Church of St. Catherine of Alexandria in Brooklyn to hear the Reverend John J. O'Neill say, "He succeeded not by politics or favor but by the force of his character and by his merit." Equitable played the mortgage on the Chief's house, and presented it to the widow, along with monies that had been sent in sympathy.

Dangerous Rescue Attempts

Three men trying to escape the flames were running wildly back and forth on the roof trying to get the rescuers' attention. The rescue ladders did not reach the roof, so a rope that the men could use to lower themselves to the ladder three stories below was shot from the National Exchange Building on Cedar Street. Unfortunately the efforts came to late. As the men were securing the rope to a chimney, the roof gave in, and one of the men plunged backwards to death in the flames, while the other two leaped to their deaths on Cedar Street. It was reported that eight men were trapped in the building, and Battalion Chief William Walsh asked for volunteers to save them. Fourteen firefighters, including Chief Walsh, stepped forward. They climbed ladders to the fourth floor, and were descending down the stairs in the search when the fourth floor collapsed and Chief Walsh was trapped. "He was on the stairway when the floor began to give way. I heard him cry 'go ahead' to the men." The firefighters survived, but heroic Chief Walsh perished. It turned out that there were three people inside the building, who were killed in addition to the three killed on the roof.

New Bern, North Carolina

DATE · DECEMBER 1, 1922

DAMAGE · 1 DEAD, $2,000,000, 600 BUILDINGS

CAUSE · UNKNOWN

The day after Thanksgiving, Friday, December 1, 1922, was cold and windy. At 10:45 a.m. an alarm was sounded for a fire at a house on Kilmarnock Street (now Craven Terrace) in the traditionally African-American

---

BELOW *Survivors look at the ruins of the predominately African-American neighborhood. To save the city, a firebreak was created by blowing up houses on Queen Street.*

community of New Bern. It took firefighters 45 minutes to arrive because they were fighting a fire across town, at the Rowland Lumber Company, the largest lumber mill in the state. The first fire had been called in at 8 a.m. and the sawmill, dry kilns and lumber shed, containing 2 million feet of board, were destroyed, causing $250,000 damage.

A series of mishaps allowed the Kilmarnock Street fire to grow out of control. After fire-

The City Before the Fire

New Bern was settled in 1710 by the Swiss under the leadership of Baron Christopher de Graffenreid. It served as the colonial capital, and in 1774 was the seat of the first provincial convention. Notable among its old buildings is Tryon Palace, the governor's mansion, which burned in 1798 and was restored in the early 1950s. During the Civil War, New Bern was occupied in 1862 by Union forces under General A. E. Burnside.

fighters arrived, having controlled the mill fire, the driver of the fire wagon discovered that there was no nozzle to attach to the hose, and he had to rush back downtown to get one. Once the nozzle had been secured it was discovered there was no wrench to turn on the hydrant and they had to go back to the station again. During this delay the fire was whipped by 45-mile-per-hour winds, sending sparks and burning debris down on the town. By the time the proper equipment arrived, three blocks of houses were burning.

Most of the houses in New Bern had wooden roof shingles, which fed the flames, and soon the fire was jumping from street to street. The fire raged for seven hours, at one point threatening the town's two hospitals. To save the city, a firebreak was created by blowing up houses on Queen Street. Other buildings were torn down with a railroad engine and heavy cable. Firefighters and volunteers, including crew of the U.S. Coast Guard Cutter Pamlico, finally got the fire under control that evening, the firebreaks allowing the remaining flames to be doused with a powerful flood of water from firefighters' hoses. Victims spent that first night under the trees in Cedar Grove Cemetery.

Hundreds of sightseers arrived by rail and water to view the ruins over the next few days. A city of chimneys marked the 40–block area, in which the fire had destroyed 600 buildings. In a controversial meeting on December 21, the city council voted to condemn about twenty acres of the burned area, to widen and straighten the streets, extend Cedar Grove Cemetery and create a park. A storm of protest followed because most of the condemned area was owned by African Americans who had lived there for years. Many families displaced by the fire moved north.

The Relief

By 10 a.m. the next morning, a relief committee was formed and tents, cots and mattresses were received from Fort Bragg. Army kitchens were set up to feed the estimated 3,000 homeless. An enormous 500–tent city was erected, with two tents per family, one for cooking and one for sleeping. One woman recalled living there for two months. Donations were received from all over the country.

BELOW *Part of the enormous tent city erected by the Red Cross.*

RIGHT *The ruins of St. Peter's A.M.E.Z. Church, at 615 Queen Street, constructed in 1914. Firefighters and volunteers, including crew of the U.S. Coast Guard Cutter Pamlico, finally got the fire under control. The church was reconstructed in 1942.*

BELOW RIGHT *Smoke and flames billow above the burning houses. Most of the houses in New Bern had wooden roof shingles, which fed the flames, and soon the fire was jumping from street to street.*

The Rebuilding

The city rallied and rebuilding began almost immediately. A new hospital was erected and many new houses were built in the burnt district. The massive effort to help the victims of the fire resulted in a new sense of pride and community for the city, which continues to the present.

"Three thousand people were left homeless, and out of this they became quite disheartened."
—Dorcas Carter, survivor

Red Cross Tent Colony

Cleveland Clinic, Cleveland, Ohio

DATE · MAY 15, 1929

DAMAGE · 124 DEAD

CAUSE · STEAM PIPE IGNITING STORED X-RAYS

An eyewitness said, "The doors were hurled open and nurses and patients fled screaming from the building…flames shot from the windows…everything was confusion…the screams of men and women rent the air." Struggling to reach safety, victims choked and died where they fell. Some of them, unclothed, were crushed beneath falling debris. Physicians and nurses fell with test tubes and charts in their hands. According to *The New York Times*, "The screams and yells of the trapped patients could be heard for blocks."

"Patients, nurses, doctors die in flight," read the headline in *The New York Times* on May 16, describing the most disastrous hospital fire in the United States, which had occurred at the Cleveland Clinic the day before. A leaky steam pipe in the basement superheated stored x-rays, causing them to explode, and in the deadly reaction, create a yellow poisonous cyanide gas from the chemicals in the photographic process.

The explosion occurred at 11:30 a.m. The flames and smoke from the exploded chemicals spread through the building quickly. Gas fumes were blown into every room, taking the lives of patients in their beds.

Firefighters, under Battalion Chief Michael Graham, were turned back by suffocating gases at every

RIGHT *Police and firemen removing the body of one of the victims. One truck driver took six loads of bodies to Mount Sinai Hospital.*

The Building Before the Fire

The 4-story building had been erected in 1920 for the founders of the clinic, Doctor George W. Crile, Doctor William E. Lower and Doctor John Phillips. There was a central atrium, from the ground floor to the roof, around which were arranged the examination rooms. At the time of the fire, it was estimated that there were 250 patients and staff present.

"The deaths were apparently due to gas poisoning. Persons in the building collapsed and were dead less than a minute after the gas was inhaled."
—Dr. William E. Lower, co-founder

entrance, and finally reached the roof by motorized ladder. They hacked their way through the skylight, and lowered themselves to the mezzanine. There they found "bodies packed deep in the space between the elevator and the stairway." They were forced backwards from their precarious position by another explosion, which shattered the skylight and rained glass and steel onto the floors below. This second explosion sent smoke and gas into the air outside, and enveloped the entire building in a hazy fog of heated poison.

Within an hour, the gas had dispersed, allowing fire-fighters and police to enter the building briefly. Volunteers helped take the injured and dead from the building. Delivery trucks, wagons and automobiles were pressed into service to transport people to hospitals. One truck driver took six loads of bodies to Mount Sinai Hospital.

There were frantic efforts at the various hospitals to save the rescued, but the poison had destroyed the membranes of the victim's lungs. Many people who had escaped died hours later from the effects of the gas.

Saturday, May 18 was declared an official day of mourning by Mayor John D. Marshall, and private burials took place over the next few days. Among the dead was one of the founders of the clinic, Doctor John Phillips.

LEFT *Doctor George W. Crile, director of the hospital, anxiously watching rescuers enter the burning building, where deadly yellow poison gas had spread.*

BELOW *The floor plan of the Cleveland Clinic. The fire began in the basement film room.*

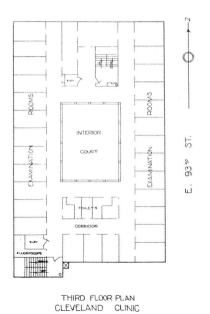

THIRD FLOOR PLAN
CLEVELAND CLINIC

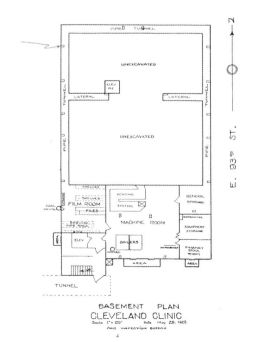

BASEMENT PLAN
CLEVELAND CLINIC

The Reaction

The *Report on the Cleveland Clinic Fire*, produced afterwards by the National Board of Fire Underwriters, blamed the leaking steam pipe for heating the film, causing the explosion. They found that the nitric and hydrogen cyanide gases produced by the heated x-ray film burned intensely when mixed with air. It recommended providing ample ventilation, fire sprinklers and proper storage of nitrate film for hospitals, doctors' offices and commercial photographic studios.

Ohio State Penitentiary, Columbus

DATE · APRIL 21, 1930
DAMAGE · 318 PRISONERS DEAD
CAUSE · ARSON

The Prison Before the Fire

The prison, housing 4,300 inmates in accommodations built for 1,500, was built in 1830. It was condemned in a 1929 report as one of the worst prisons in the country, for its inhuman conditions.

Over 300 prisoners were burned to death in their cells when a fire transformed the Ohio State Penitentiary into a human pyre.

At about 6 p.m., when all the prisoners had been locked in their cells for the night, a small blaze was discovered in the northeast section, on the roof of blocks G and H, later to be revealed as an act of arson. The fire was aided by a strong wind, and spread rapidly to the I and K blocks.

RIGHT *Prisoners clear away debris where flames swept through dormitories and part of the main cell blocks, killing 318 of their prisonmates.*

A Prisoner Hero

Prisoners worked in the flames to free their fellow inmates. Holward Jones, a prisoner, proved a hero when he seized a sledgehammer and single-handedly rescued 136 convicts by hammering the locks off their doors. According to other reports "William Wade Warren...prisoner, broke down a cell door with a sledgehammer and released twenty-five men." Another prisoner seized keys from a guard and released many prisoners. Big Jim Morton, a notorious Cleveland bank robber, rescued scores of men from K cell block.

As the fire gained headway many guards in the units would not free the prisoners, for fear of a mass escape. One guard in particular, Thomas Watkinson, refused to hand over his keys until other, more humane guards finally wrestled him to the ground. They opened as many cells as they could before smoke forced them back. Glenn Pierce, one of the prisoners, declared, "The lives of 100 men would have been saved had the guards consented to open the cells." Virtually every man on 5H knew he would die because the rescue parties, cut off by the flames, could only reach 3H.

> "Ohio had ignored for more than ten years official evidence that the State Penitentiary at Columbus was apt any day to be a firetrap."
> —William H. Allen, Director of the Institute for Public Service

OPPOSITE *Firemen fighting the prison fire in the building, which was built in 1830, prior to the advent of fireproof construction, such as hollow tiles and steel-reinforced walls.*

RIGHT *Warden Preston E. Thomas testifying about the crammed conditions, lack of fire-fighting equipment and reluctance of guards to open cells. The government inquiry into the incident placed blame on Thomas for not releasing the prisoners earlier, resulting in many deaths. However, he was not removed from his post until the prisoners, housed in tents after the fire, revolted.*

In the meantime, Warden Preston E. Thomas, who was more concerned about a breakout than the lives of the prisoners, had gone outside the prison to direct guard operations. On the outside lawns, around the perimeter of the prison, prison guards, National Guardsmen and regular Army troops were positioned to stop any prisoners from leaving. However, there were no attempts to escape.

By midnight the fire had been controlled, and the dead were being removed to a temporary morgue at the state fairground, while the 150 injured were hospitalized.

Still under Warden Thomas's direction, the prisoners were housed in the yard in white tents that soon became known as "White City." Protests by prisoners over conditions brought in more troops. On April 28 convicts started a new revolt, and the National Guard fired tear-gas bombs into the crowds, while pursuing some convicts with fixed bayonets. The next day two rebelling prisoners were shot and wounded by rifles when a crowd of 1,300 prisoners staged an uprising. On May 7, prisoners set fire to 50 army tents in the "White City." Authorities said that prisoners who had burned their tents would have to sleep in the open air because the tents would not be replaced. Many found shelter in improvised pup tents made out of sections of burned tents, blankets and sheets. On May 8, machine gun fire killed two convicts while they slept on the ground in the yard. A guard claimed it accidentally discharged, but the situation worsened until Governor Cooper suspended Warden Thomas.

The Investigation

An inquiry called by Governor Myers Cooper placed the blame for the delay in releasing the prisoners on Warden Thomas, but Governor Cooper refused to suspend him, saying "the authorities of the institution must be upheld." Within days, the prisoners were demanding the ouster of the Warden and threatening organized resistance. Warden Thomas, in turn, blamed the state for not allocating more funds for prisons, and was quoted as saying "thus the overcrowding." Over the summer a convict said he knew the men who had set the fire, and told the Deputy State Fire Marshal Joseph Clear. The allegations were investigated, and on August 15 two prisoners were secretly arrested for setting the April fire and were taken to the Columbus jail. They were held for murder in the first degree for the deaths caused by their arson.

The Reaction

Governor Cooper found additional executive funds to build more dormitories at other prisons, and the conditions at the Ohio State Penitentiary gradually improved. The fire prompted numerous editorials deploring the prison system, and focused attention on the physical condition of prisons nationwide. Many prisons built prior to 1850 were closed and fireproofing was put in place. The worst fire in U.S. prison history resulted in improved conditions for many prisoners.

Reichstag, Berlin

DATE · FEBRUARY 28, 1933
DAMAGE · DESTROYED
CAUSE · ARSON

"Incendiary fire wrecks Reichstag; 100 Red Members ordered seized."
—*The New York Times* headline
February 28, 1933

The Building Before the Fire

The Reichstag building, fronting on the Koenigsplatz, was built as the seat of government between 1884 and 1894 at a cost of more than $5,000,000. It was in the Italian Renaissance style and was designed by Paul Wallott. It was funded out of the indemnity extracted from the French after the Franco-Prussian War of 1870. Built of Silesian sandstone, it was 430 feet long with a central glass dome 246 feet high supported by gilded girders. The main session hall occupied the area under the great dome.

The fire was discovered shortly after 9 p.m. when a patrolling policeman smelled smoke. He hurried towards the parliamentary chamber, and was met by a suspicious group of men who failed to halt on his demand. He managed to capture one of them, an alleged Communist named Marinus van der Lubbe. Other policemen rushed to the chamber to find rugs and chairs piled together and ablaze. The alarm was sounded, but by the time the firefighters arrived, the flames were spreading rapidly. The chamber was paneled in wood, and the

RIGHT *Smoke billows from the glass dome, which was 246 feet high. Note the ladder on the right, which firemen used to climb to the roof in a futile effort to fight the fire, which had suspiciously started in many different places in the building.*

BELOW *While the fire completely destroyed the interior of the main chamber and dome within one-and-a-half hours, firemen managed to save the library and archives. Inspection of the charred support beams revealed that they broke in half, pulling away from the walls and collapsing the dome.*

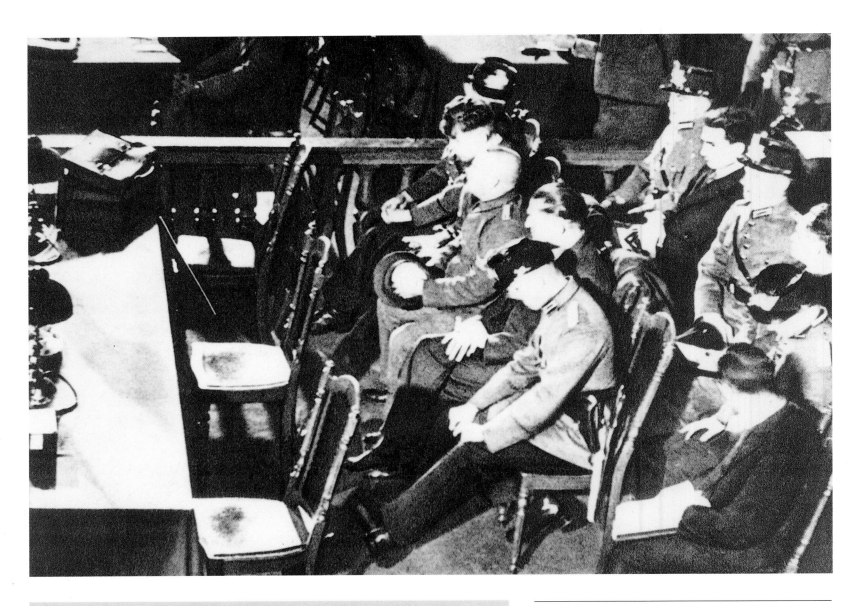

Charging the Opposition

Chancellor Hitler, Prime Minister Hermann Goering and Joseph Goebbels inspected the damage that night. Early the next morning, Goering ordered the arrest of 100 Communist members of the Reichstag. The government's activities, aimed at silencing all opposition, immediately took on an increased pace. The Communist newspapers were ordered closed, their headquarters shuttered and meetings were prohibited. The Cabinet approved a law imposing the death penalty for treasonous activities. This included supplying news to the foreign press deemed "anti-German."

Additional Communists were held for trial. During the trials, the judges, witnesses and prisoners toured the building, in a reconstruction of events on the night of the fire.

Following the fire, Hitler was granted dictatorial powers under the Enabling Act and a state of emergency was declared throughout the country. His election as Führer the following year precipitated World War II.

ABOVE *The Reichstag fire trial. The accused— Dimitrov, Torgler, Popoff and Tanev—were acquitted. Van der Lubbe was found guilty and guillotined.*

benches, desks and chairs were also wooden, and the blaze soon made the chamber a veritable furnace.

The intense heat caused the glass dome to fall down quickly and flames spouted through the roof. By 10:30 the whole structure seemed doomed. Firefighters concentrated on saving the library and records room. Under their efforts the fire was contained to the main chamber, and by 11 p.m. only a few smoldering embers remained.

For many years it was believed that Nazi agents had started the Reichstag fire as a means of launching a campaign against the Communists, and seizing power. The debate over who set the fire continues today.

Morro Castle Ship

DATE · SEPTEMBER 8, 1934
DAMAGE · 130 DEAD
CAUSE · UNKNOWN

The disaster occurred in heavy rain and gale-force winds, about eight miles off the New Jersey coastal resort of Asbury Park. Captain Robert Willmott died at about 8:45, of a heart attack, just eight hours before the SOS was sent by George Rogers, Chief Radio Operator, at 4:23 a.m. Chief Officer William F. Warms was in command when fire broke out in the midship section and spread rapidly. Although the ship was equipped with the latest fire-fighting technology, the thick smoke in the engine room, combined with highly flammable polish material stored where the fire broke out, overcame fire-fighting efforts. In addition, it was discovered later that the crew did not man their stations, instead fleeing to save themselves. The flames climbed from A and B decks upward, shooting fifty to sixty feet in the air. Many passengers were trapped in their cabins, unable to escape through the burning hallways.

The six lifeboats on the port side were inaccessible because of the fire, leaving only four that were launched. Some passengers later charged that crew members crowded onto the

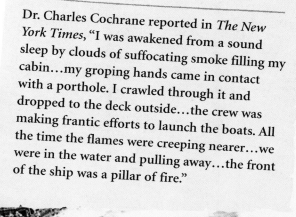

Dr. Charles Cochrane reported in *The New York Times*, "I was awakened from a sound sleep by clouds of suffocating smoke filling my cabin...my groping hands came in contact with a porthole. I crawled through it and dropped to the deck outside...the crew was making frantic efforts to launch the boats. All the time the flames were creeping nearer...we were in the water and pulling away...the front of the ship was a pillar of fire."

RIGHT *People stand on the beach at Asbury Park and watch smoke pour out the Morro Castle, a day after it caught fire. Boat fires differ from land fires because there are no fire departments to call to the scene. They must be fought from onboard and crewmen are often ill-trained and unprepared for a large fire.*

The Ship Before the Fire

Cuba in the pre-Castro era was a wild, fun-filled tourist destination, with large hotels, casinos and stage shows. Many luxury cruise ships plied the waters between New York and Havana. The *Morro Castle*, named for the fortress in Puerto Rico, was one of the Ward Line's ships that regularly ran tourists between the two cities. It had been designed by Naval architect Theodore F. Ferris of New York, and launched in 1930. It cost about $4,000,000, was 508 feet in length, 70.9 feet at the beam and 39 feet at the draft. After a Labor Day holiday cruise, the ship sailed from Havana harbor for the return trip to New York, and carried 318 passengers, 230 crew and ten officers.

boats, leaving the passengers to burn to death or jump into the dangerous sea. The Chief Engineer, Eben Abbott, was in Lifeboat 1 with three passengers and twenty-nine crewmen, while Lifeboat 3 got away with sixteen crew but no passengers, when it could carry 70. Frightened passengers in the aft were cut off from the rest of the ship, and many jumped overboard into the pounding surf.

The boardwalks at the resorts became jammed as crowds gathered to watch the fire. At 10 a.m. Captain Warms abandoned the bridge, once two lines were secured to the still-burning ship. Many survivors told of clinging to ropes for six hours, or of swimming for hours until rescued. Couples who had leaped together became separated after hours in the water, and in many cases only one survived, while the other drowned.

New Jersey Governor A. Harry Moore participated in the rescue work by boarding a small plane, flying to the site and waving a red flag to point out survivors in

Rescuers

The passenger ships *Monarch of Bermuda* and *City of Savannah* turned around in response to the SOS, and arrived about 7:30 a.m. The freighter *Andrea S. Luckenbach* arrived about 6 a.m., along with some Coast Guard boats. Captain A. R. Francis of the *Monarch of Bermuda* launched four power boats into the choppy seas, saving 71 people. Passengers aboard the *Monarch* described seeing a sailor with a child on his back dive into the water and swim to a boat.

the water to the rescue boats, which had been having trouble spotting people in the high waves.

On Sunday evening, September 9, the smoking liner was driven ashore directly off the Asbury Park Convention Center Pier, when the Coast Guard's attempt to tow it failed. It was beached broadside, with its bow pointing north. City officials claimed it as a tourist site and charged 25 cents admission to the Convention Center for viewing. Lines of automobiles carrying spectators created the worst traffic jam of the summer.

Survivors formed the Morro Castle Association, and condemned the inquiry report that placed the blame for the fire on a few of the officers, rather than on the Ward Line itself, for allowing lax conditions onboard. More than 200 claims were filed for a total of nearly $2,000,000.

LEFT *The smoldering hulk beached off the board-walk at Asbury Park shows that the structural steel survived, but the interiors were gutted due to the inflammable construction materials.*

BELOW *Rescuers pull victims from the wooden lifeboats of the burning* Morro Castle.

Crystal Palace, London

DATE · NOVEMBER 30, 1936
DAMAGE · DESTROYED
CAUSE · UNKNOWN

Engulfed in a towering sheet of flame, which roared so high into the night sky that it could be seen from the English Channel, the world-famous Crystal Palace, architectural pride of the Victorian era, crashed to the earth, a raging inferno of twisted girders and molten glass.

ABOVE *Firemen shoot water from a truss-type, extension aerial ladder to contain the fire. These ladders were adjustable to reach different heights. They replaced older wooden ones and have now been supplanted by aluminum and fiberglass models.*

The Building Before the Fire

The Crystal Palace was designed by Sir Joseph Paxton, for the Great Exhibition of 1851. Originally located in Hyde Park, it was opened on May 1, 1851, with great pomp and ceremony, by Queen Victoria and Prince Albert, who had organized the international exhibition. It covered 755,208 square feet, which was seven times the area of St. Paul's Cathedral. It was constructed of prefabricated wrought-iron elements, holding 25 acres of glass panels. It was 1,848 feet long and 408 feet wide. The central nave was 72 feet wide. There were 14,000 exhibitors, with more than 8 miles of display tables. A year later, it was moved piece by piece to Sydenham Hill, on the southern side of London. It was used over the next 75 years for concerts, circuses and special exhibitions.

The Ruins

Within three hours of the start of the fire, the Crystal Palace was a smoldering, charred ruin. Several firefighters were injured fighting the flames. The two towers which survived the fire were demolished in 1941 because they were considered a target for German bombers.

ABOVE *Melted glass from the Crystal Palace fire, on display at the Natural History Museum in South Kensington, London. Glass melts at 1563° F.*

LEFT *Flames shoot hundreds of feet from the Crystal Palace. The cast–iron girders melted at 1500°F. One of the two towers that survived rises up from the smoke. They were demolished before WWII, for fear that they might be bombing targets.*

The fire, of unknown origin, was discovered shortly after 8 p.m., and firefighters arrived from all over London. However, efforts to check the flames were futile, and within minutes, the great central hall, or nave, was a mass of heavy, billowing smoke and flames. The wooden floors, as well as the wooden exhibit stalls, fed the flames. By 8:30 p.m. the domed arcade of glass, towering 175 feet, collapsed, sending up showers of sparks and blazing embers. Great tongues of flame shot into the sky, as people gathered to watch the spectacle. Shortly after 9 p.m. the large quantity of fireworks stored there had exploded, which was the beginning of the end of the structure.

Although there were many persons in the building when the fire began, they all escaped uninjured, due to the proper placement and number of exits. One thoughtful person threw open the cages in the aviary, releasing thousands of birds that fluttered through the smoke to the trees outside.

Mounted police and patrolmen fought in vain to control the thousands of awestruck spectators who pressed closer. One of the two 300-foot-tall towers caught fire and nearly collapsed on the crowd, but was extinguished by firefighters first.

The Times of London editorialized, "It stood long enough in its enlarged form to outlive, perhaps, the last of all such exhibitions…. Its destruction is like that of a cherished and venerated historical document."

Hindenburg Airship

DATE · MAY 6, 1937
DAMAGE · 35 DEAD
CAUSE · UNKNOWN

The Airship Before the Fire

The *Hindenburg*, originally designated the LZ129, was the largest aircraft ever built. It was 803.8 feet long with a height of 146.7 feet. Its diameter was 135.2 feet and it weighed 430,850 pounds. The hydrogen gas capacity was 7,602,100 cubic feet.

The hydrogen-filled German dirigible *Hindenburg* had been due to arrive at the Lakehurst Naval Air Station in New Jersey at 6 a.m., but was delayed by strong headwinds over the Atlantic Ocean. Shortly after 3 p.m. it flew over New York City, providing thrilled passengers a stunning view of the Empire State building. The airship was on the last leg of its first transatlantic voyage of the year. At Lakehurst with the dirigible only 260 feet above ground, passengers could be seen in the windows laughing and waving to the people on the ground. The mooring rope was dropped from the starboard bow.

At 7:23 witnesses heard an explosion and suddenly the ship burst into flames just above the rear fin. The 92-man naval ground crew, and 139 civilians, ran for their lives, as the stern of the ship sank to the ground, while the fire spread upwards toward the nose of the ship. A blast of flame shot into the air from midship,

> "It burst into flames!…This is one of the worst catastrophes in the world!… Oh, the humanity and all the passengers!"
> —Herbert Morrison's famous radio broadcast

and the silver fabric skin burned away in fifteen or twenty seconds. The exposed metal framework crashed onto the landing area in a mass of red flames and black smoke. Only 34 seconds had elapsed between the first explosion and the final wreckage.

Frantic passengers waited as long as possible before leaping from the burning cabin through the windows or climbing down the mooring ropes to the ground. One woman threw her three children out the window before leaping. Several survivors reported being blown out of the windows by the backdraft of the fire. Some were in flames as they plunged to earth. Others who were not near the windows crawled out of the wreckage once it hit the ground, their clothing burned and their bodies seared. Screams of the injured and dying spurred rescuers on.

Firefighters fought the flames for several hours, while rescuers braved the heat to drag out injured persons and the bodies of the dead. Ambulances transported

The Escapees

Amazingly, 25 passengers and 44 crew members escaped the burning ship, including Captain Max Pruss, commanding officer of the airship, and Captain Ernst Lehmann, who both leaped from the command gondola below the framework. Captain Lehmann, who had commanded the *Hindenburg* from its first flight in 1936, died the next day of his burns. His last words were "I am going to live."

LEFT Hindenburg Airship *on June 22, 1936, on a previous trip to Lakehurst after a flight from Frankfurt.*

ABOVE *The German airship exploded just prior to landing at Lakehurst, possibly due to a spark from static electricity igniting the flammable hydrogen gas.*

the injured to hospitals and the dead to an improvised morgue on the naval reserve. Many of the bodies were charred beyond recognition and were laid out in rows for identification. Among the dead was one member of the ground crew.

Thousands of sightseers clogged the highways, hoping to view the wreckage. Army guards were posted every 100 feet to prevent intruders.

The burning of the *Hindenburg* marked the beginning of the end of the era of large, lighter-than-air zeppelins. The advent of airplane passenger service dealt a deathblow to passenger flights by airships.

The Investigation

A Federal investigation board, a U.S. Navy board and a German board investigated the crash. Most scientists attributed the fire to the highly inflammable hydrogen gas, ignited by a spark of static electricity. They contrasted it to the non-explosive helium gas used in U.S. airships. Professor Otto Stern of Carnegie Institute of Technology in Pittsburgh attributed the cause to a "demon proton" in the hydrogen atom. Hitler called it "an act of God."

However, one of the popular theories of today is that the ship was sabotaged by someone who brought a bomb onboard. Theorists believe it may have been an attempt to embarrass Hitler and the Third Reich. Whether the bomber was a crewmember or a passenger is a matter of speculation.

Dance Hall, Natchez, Mississippi

DATE · APRIL 23, 1940
DAMAGE · 198 DEAD
CAUSE · UNKNOWN

The Dancehall Before the Fire

The hall was decorated with Spanish moss draped over the rafters to conjure up a rustic look. The corrugated iron building, which had formerly been a blacksmith shop, had just one front door and all the windows had been boarded up to prevent gatecrashers.

Three hundred people from the African-American community of Natchez were gathered at the Rhythm Night Club on St. Catherine Street for a Tuesday night dance.

The Walter Barnes Orchestra of Chicago was playing when the fire broke out near the front door. A handful of dancers battled their way through the narrow door or wormed their way through the ticket-sellers window nearby. The rest were driven toward the rear of the building, blinded by smoke and flames that quickly crackled through the masses of dry Spanish moss.

The stampede swept back around the bandstand, beating futilely against the securely boarded windows. One rear window gave way and a few struggled out. The rest died screaming and clawing for escape and fell in piles that mounted shoulder high near the bandstand.

Fifteen minutes after it began, the fire was under control, and firemen fought their way inside. Moans came from beneath the mounds of scorched bodies, and those who were still alive were brought out. Twenty of these burned and battered victims died on the way to the hospital. Doctors said that most of the dead had been suffocated by thick smoke or crushed to death in the terrible stampede, rather than killed by the flames.

RIGHT The exterior of the Rhythm Club following the fire, which heated up the corrugated iron building like an oven. Temperatures inside may have reached over 1,000°. There was only one exit.

The Investigation

An immediate investigation resulted in the arrest of five men who had been heard threatening to set the building on fire. They were released due to lack of evidence. The cause of the fire was never determined, although it is believed to have started near the front door by a careless smoker.

The Blitz, London

DATE · SEPTEMBER 1940–
MAY 1941
DAMAGE · 19,566 DEAD,
25,578 INJURED
CAUSE · INCENDIARY BOMBS

"If the British Empire and its Commmonwealth last for a thousand years, men will still say, 'This was their finest hour.'"

—Prime Minister Winston Churchill

German aircraft made their first major raid on England the night of June 18–19, 1940. Bombs fell in East Anglia and in Surrey. Throughout June, various raids on industrial targets were carried out by the Luftwaffe. As the Battle of Britain grew over the summer, night raids became extensive.

OPPOSITE *Firefighters use a pumper truck to fight fires on Victoria Street caused by the bombing. The pressure generated by the engine could shoot water 80 feet in the air. Hundreds of fires were valiantly fought each night by a volunteer force of men and women.*

BELOW *The dome of St. Paul's Cathedral silhouetted against the flame-lit sky, after surviving German bombs on June 7, 1941.*

One spectator wrote, "St. Paul's during this night provided the most inspiring and terrifying sight that Londoners had ever seen…the dome seemed to ride the waves like a galleon."

LEFT *Firemen direct their hoses onto smoking buildings devastated by an air raid. In many parts of the city, firefighters had to bring in water-filled trucks because of broken water mains, but fallen debris frequently prevented the trucks from getting close enough to the building to battle the fire.*

The first bombing of London came on the night of September 7, 1940. The attacks were aimed at weakening the morale of the British people, possibly showing the futility of resisting and forcing the British government to sue for peace. They were also viewed by the Germans as reprisals for attacks on German civilians. Over the first week, the East End of London was heavily damaged.

During September and October most of the damage was caused by incendiary bombs. Historian John Ray provides some figures on the bombings: In December the Germans dropped over 600 tons of bombs and 4,000 canisters of incendiaries. On December 8 alone, there were 115,000 fire bombs. The next morning it was reported that "there were 9 major, 24 serious, 202 medium and 1,489 smaller fires."

There was no bombing over Christmas, but on December 29, the Germans dropped 22,000 incendiary bombs. The fire built into an inferno, which destroyed eight Christopher Wren churches and numerous houses.

The temperatures reached 1,800 degrees fahrenheit in some locations. Some of the fires were still burning days later. Eight firemen were killed near Fleet Street when a burning wall collapsed on them. The number of people killed in December was 3,793, with over 5,000 injured.

Winter weather slowed the raids in January and February, but from March to May London continued to be attacked. There were major raids on March 8 and 9, as well as on the 19th, when over 120,000 incendiaries rained down on London. The last large raid was on May 10, when over 2,000 fires were started, burning portions of Westminster Abbey, the Houses of Parliament, the Tower of London and 5,000 houses.

The Preparations for War

Since the mid-1930s, Britain had been preparing its citizens for the possibility of war, while hoping for a peaceful solution. The British government strengthened the Royal Airforce with the introduction of light bombers and faster fighter planes. An Air Raid Precautions Department had been formed, which began to address practical issues such as blackouts, shelter, food and sanitation. A voluntary Air Raid Wardens' Service was in place by 1937, and in 1938 the Auxiliary Fire Service was established. Britain and France declared war on Germany on September 3, 1939, following the invasion of Poland by Germany on September 1. Churchill became Prime Minister on May 10, 1940, and that day the German *Blitzkrieg* began. Soon Holland, Belgium, Luxembourg and France fell.

Pearl Harbor, Hawaii

DATE · DECEMBER 7, 1941
DAMAGE · 2,300 DEAD
CAUSE · MILITARY ASSAULT

"A date which will live in infamy."
—President Franklin D. Roosevelt

Under a plan by Admiral Isoroku Yamamoto, Commander-in-Chief of the Japanese Combined Fleet, six aircraft carriers, two battleships, three cruisers and eleven destroyers were stationed at a point about 275 miles north of Hawaii. Yamamoto planned to destroy the U.S. Fleet, which would clear the way for Japan to invade all of Southeast Asia.

Sunday, December 7, 1941 was the date of the surprise aerial attack on the U.S. Pacific Fleet naval base.

The first Japanese dive bomber appeared at 7:55 a.m. Hawaiian time and was quickly followed by 353 bombers, torpedo planes and fighters. At 8:10 bombs ignited the forward magazines on the Battleship USS Arizona, resulting in a tremendous explosion and fireball that sank the ship in nine minutes, claiming the lives of more than 1,100 men. The U.S.S. Oklahoma capsized and the U.S.S. California, U.S.S. Nevada and U.S.S. West Virginia sank in shallow water. An explosion rocked the Destroyer U.S.S. Shaw. Three other battleships,

The Harbor Before the Bombing

Pearl Harbor, named after the pearl oysters which once grew there, is located on Oahu Island, and had a harbor area of more than 10,000 acres, with hundreds of anchorages. Approximately 100 ships of the U.S. Navy were anchored in the calm waters that morning, including the battleships U.S.S. Arizona, U.S.S. West Virginia, U.S.S. California, USS Oklahoma, U.S.S. Nevada, U.S.S. Pennsylvania, U.S.S. Tennessee and U.S.S. Maryland. The light cruisers included the U.S.S. Detroit and the U.S.S. Raleigh. The destroyers U.S.S. Cassin and U.S.S. Downes were in drydock 1 and the destroyer U.S.S. Shaw was in drydock 2. There were also smaller service ships present.

BELOW *Explosions and flames on Battleships row. Eight battleships, three cruisers, three destroyers and several other vessels were lost or damaged.*

BOTTOM *Thick smoke and flames pour from the battleship U.S.S. Arizona which sank within 9 minutes, with a loss of 1,100 men. A white concrete and steel structure now spans the hull of the sunken ship, which was dedicated as a national memorial on May 30, 1962.*

three cruisers, three destroyers and other vessels were damaged, and hundreds of aircraft were destroyed.

Simultaneously, Japanese bombers attacked nearby Hickam Field, Wheeler Field and the Naval Air Station on Ford Island, where U.S. fighter planes had been lined up in rows on the airfields and made perfect targets. At 8:26 the Honolulu Fire Department responded to a call from Hickam Field, but the fires were already burning beyond control and enemy aircraft was strafing the field with gunfire. Three firemen were killed and six were injured trying to put out fires.

At 8:54 the second attack run by Japanese bombers began, and at 9 a.m. the crew of the Dutch ship Jagersfontein responded with her guns, the first Allies to join the fight. At 9:30 the destroyer U.S.S. Shaw exploded, sending debris everywhere.

The surprise attack temporarily severely crippled the U.S. naval forces, but six of the damaged battleships were repaired and put back into service. Some of the Pacific Fleet had been out to sea, including the fleet's three aircraft carriers, and so escaped the destruction.

After the war there were questions of how much advance information U.S. officials possessed, and whether it was purposely withheld to bring the U.S. into the war.

A National Memorial was dedicated on May 30, 1962, which spans the hull of the sunken U.S.S. Arizona. In 1991, Fiftieth Anniversary activities took place on the Memorial, with hundreds of survivors attending.

RIGHT *Firemen battle flames on the U.S.S. West Virginia from fireboats. The operating principle of fireboats has not changed much over the past century—they pump water up from the bay or river where they float and shoot it out powerful water canons onto the fire.*

The Victims

The majority of damage to ships and planes occurred during the first 30 minutes of the attack. U.S. casualties totaled more than 2,300 dead with over 1,000 injured. The Japanese lost 29 aircraft.

The Reaction

The attack united the American people and precipitated the entry of the U.S. into WWII. However, the U.S. military was criticized for its lack of defense and poor preparation. Official investigations began immediately, and Admiral Husband Kimmel and General Walter Short were relieved of their commands.

S.S. Normandie

DATE · FEBRUARY 9, 1942

DAMAGE · 1 DEAD

CAUSE · SPARK FROM A CUTTING TORCH

The French luxury liner was built from 1932–1935 at St. Nazaire at a cost of $56,000,000. The 83,423–ton ship made its bid for the mythical "Blue Ribbon of the Atlantic" speed record on her maiden voyage from Le Havre to New York. She arrived on June 3, 1935 in the record time of 4 days, 11 hours and 33 minutes, an average speed of 29.68 knots. Mademoiselle Albert Lebrun, wife of the French President, and "godmother" of the ship, was a passenger on the maiden voyage.

The Normandie departed Le Havre for her last voyage on August 23, 1939, and docked on the 28th at Pier 88 in New York. Due to the impending hostilities of World War II, it was decided to lay up the ship. France and Great Britain declared war on Germany on September 3, 1939 and the Normandie remained at berth with a reduced crew of 113.

According to author Charles Offret, "In April 1941, the press made a public issue of the suspicion that the Normandie's crew might be preparing to sabotage their ship." Captain Le Huédé vehemently denied it, but on May 15, 1941, the U.S. Coast Guard occupied the ship and placed it under protective custody. On December 7 the Japanese attacked Pearl Harbor, and on December 11 war was declared on Germany and Italy. On December 12 all French ships were seized, and the crews removed. On January 1, 1942 the ship was renamed the Lafayette, and preparations began to convert her into a troop ship. Most of the luxurious fittings and furnishings, including silverware and wine, were removed. Numerous delays prolonged the conversion, and hundreds of workmen were brought aboard to convert the ship by the sailing deadline of February 14.

On February 9, workers were using acetylene torches to cut down four tall metal lighting towers, weighing 500 pounds each, in the Grand Lounge. The lounge was being used for storing over 1,000 bales of life jackets filled with kapok, material that can support 30 times its weight in water, but that is also highly flammable. It was also used for stuffing pillows, mattresses and upholstery, among other things. Charles Collins, a workman, told *The New York Times*, "A spark hit one of the bales and the fire began. We tried to beat out the fire with our hands…the smoke and heat were just ter-

The Ship Before the Fire

With an overall length of 1,029 feet, the Normandie had a passenger capacity of 2,170, including 930 first class, 680 tourist and 560 third class. In the beautifully illustrated book, *Normandie: Queen of the Seas*, by Charles Offret, an essay by Bruno Foucart on "Art on Board Normandie" discusses the sumptuous interiors: "teams of well-known, proven artists came together, therefore, and defined a style that would be as much *transat* as *art deco*." There were huge glass panels by Dupas in the Grand Lounge, depicting the mythology of the sea. There was furniture by Ruhlmann, lacquer panels by Dunand, done in an "Egyptian" style for the Smoking Room. A sculpture by Baudry, personifying the Normandie, stood at the top of the stairway to the Grill Room. There were bas-reliefs by the Martel Brothers, cabinets by Leleu, enormous bronze doors by Subes, interiors by Patou and Pacon, and huge crystal chandeliers by Lalique. The gracious grand foyer, three decks high, featured walls of Algerian onyx. Many famous celebrities crossed on the Normandie, including Josephine Baker in 1935, Marlene Dietrich and Cary Grant in 1938, and Jimmy Stewart and Bob Hope in 1939.

> "There is no evidence of sabotage. Carelessness has served the enemy with equal effectiveness."
>
> —Frank S. Hogan, District Attorney

The Victim

Frank Trentacosta, a civilian workman and member of the fire watch, was killed when a feeder tank for the torches exploded. Many of the firefighters who got close to the flames with hoses said that the terrific heat sent the hose spray back at them in the form of live steam, which stung and reddened the skin.

ABOVE *Firemen in the flame-swept ballroom. Note the light torchere—it was the cutting down of one of these which caused the fire. So much water was pumped into the ship that shortly after this photo was taken she capsized onto her side.*

rific." The fire began at 2:35 p.m. and firemen arrived fifteen minutes later, while the crew and workers were abandoning ship.

Firemen on the pier and in fireboats poured water on the flames, which roared through the three upper decks. Flames could be seen at openings in the bridge. All afternoon, in the darkness created by the fire's smoke, firemen from Hook and Ladder Companies 2 and 35, stationed on the West Side Highway, carried victims from the ship on stretchers.

Many men climbed down an 85–foot firemen's ladder extended from the West Side Highway to the bow. Heavy smoke drifted over midtown, drawing hundreds of spectators to the scene.

Francis Dieck, a shipfitter's helper told *The New York Times*, "We formed a human chain and felt our way through the smoke that kept getting thicker by the minute. We groped our way up three decks to the air and then we took a hand with the hose until we were almost overcome."

By 4:45 there was over 5,000 tons of accumulated water in the ship, from all the water that had been pumped by firefighters and fireboats. The ship began to list to port about 15 degrees. By 8 p.m., when firemen gained control and the flames were extinguished, the list had increased. The portholes had not been closed, and at high tide, around midnight, more water poured into the burnt hull. Shortly before 3 a.m., as searchlights played on the once-beautiful ship and the crowd

gasped, she capsized and lay on her side in fifty to sixty feet of water and mud, at an angle of 80 degrees. The three great funnels lay just clear of the icy water.

Normandie lay for months, until pumping was begun on June 11. However, it was not until a year later, in September 1943, that the ship was upright. Stripped of her funnels, she was towed to the Brooklyn Navy Yard to determine if she could be refurbished. It was decided that she was not worth salvaging. The Normandie was sold for $161,680 on October 3, 1943 to a scrap metal dealer, and towed to Hoboken for dismantling.

It was an ignominious ending for the ship, which, barely seven years before, had been praised as the greatest and most luxurious ship ever created.

BELOW *The Normandie after it had been stripped of its funnels and decorations, being towed down the New York bay. It was later sold to a scrap dealer.*

Cocoanut Grove, Boston

DATE · NOVEMBER 28, 1942
DAMAGE · 487 DEAD
CAUSE · UNKNOWN

Downstairs in the Melody Lounge, a 16–year-old bus-boy, Stanley F. Tomaszewski, struck a match near an imitation palm tree so he could replace a light bulb in the coconut shell light. About that time, fire broke out

THE FIRE-FIGHTING FORCE

The Boston Fire Department sent out 25 engine companies, five ladder companies, one rescue company, a water tower and other trucks.

The Nightclub Before the Fire

Holy Cross had defeated Boston College that afternoon in the last football game of the season, with a score of 55–12. Jubilant fans were crowded into the bars of Boston, and patrons stood three and four deep at the bar of the Melody Lounge in the basement of the Cocoanut Grove nightclub. The lounge was down a single flight of stairs from the foyer, with an oval bar in the center and a zebra-striped banquette circling the wall. Beyond the upstairs foyer was the Grove's main dining room, which was surrounded by The Terrace, an elevated VIP section, as well as the elevated "Villa", designed in a Spanish style. Above the dance floor was a roof, which could be rolled back electrically for dancing under the stars. The stage show was to begin at 10:15 p.m., and the room was packed with patrons, including the Western star Buck Jones.

and after some futile attempts to extinguish it, the crowd of nearly two hundred people were clutching wildly at each other trying to climb the single staircase. The lounge's walls were decorated with rattan and the ceiling was blue satin, which fueled the flames.

Because the lounge was physically isolated from the main part of the club, the fire remained secret until an employee ran up the stairs. The main door to the Cocoanut Grove was a single revolving door, which quickly became jammed as the wild horde pushed forward. Suddenly a massive ball of flame burst into the main dining room, extinguishing the lights and sending patrons fleeing. The flames lit the satin ceiling and the leatherette seating, and a sweet smelling gas—fumes from the burning fabrics—was descending on the crowd, gently lulling them to sleep.

The third and fourth alarms had brought fire apparatus and equipment from every part of the city. By 10:30,

LEFT *Exterior of the Cocoanut Grove following the fire. There was only one revolving door to the club, in the center bay, under the sign. It jammed, trapping numerous victims. Fire laws were revised to mandate increased numbers of exits, exit signs and sprinklers.*

OPPOSITE *Interior of the Cocoanut Grove after the fire. The columns were wrapped with rattan to resemble coconut palm trees, which helped fuel the fire. Many victims were suffocated by poisonous gases from the burning satin ceiling fabric, still visible hanging in tatters, and the leatherette seating.*

"I wish I had died there with the others."
—Barney Welansky, Cocoanut Grove owner, on his release from prison, November 7, 1946

when the mayor of Boston, Maurice Tobin, and his fire commissioner, William Arthur Reilly, got to the scene, the department was waging a winning battle on the blaze. Holes were broken in walls to try to save people, but by 10:45 even the moans and whimpers had died out.

It had taken just 7 minutes to destroy the Cocoanut Grove. Firemen began piling the bodies in makeshift morgues in a garage and a publishing house. Newspaper trucks, taxis and cars were commandeered to move the injured to hospitals. Victims were brought into Boston City Hospital at the rate of one every 11 seconds. More blood plasma was used than was needed for American servicemen wounded at Pearl Harbor. It took 175 men five hours to remove all the victims.

A New Theory

Fifty years after the fire, in 1992, a new theory emerged when an air conditioning repairman named Walter Hixenbaugh read an article describing the unknown cause of the fire. He believes that methal chloride had been substituted for freon in the air conditioning system to save money during the war. He thinks that the sweet-smelling gas which was reported to have killed people may have leaked from the cooling system. No official investigation has confirmed this, and the cooling system no longer exists.

Ringling Brothers Barnum & Bailey Circus, Hartford, Connecticut

DATE · JULY 6, 1944
DAMAGE · 167 DEAD
CAUSE · UNKNOWN

Despite the 90 degree temperature, the Ringling Brothers Barnum & Bailey Circus Friday afternoon matinee was jammed with an estimated 8,000 people, mostly women and children. Alfred Court's animal act had just finished and the third act, the Flying Wallendas, was preparing for their performance high above the heads of the excited children. Spotlights outlined them against the roof of the big top, when a tiny flame was seen on the sidewall of section A. Ushers

ABOVE *The tent collapses as patrons flee. Pieces of burning canvas that had been coated in a mixture of gasoline and paraffin for waterproofing fell onto the heads of victims, igniting their hair and clothes.*

rushed with buckets of water to extinguish it, but before they reached it, it leaped up the canvas wall towards the roof. At 2:40 p.m. the band began to play "Stars and Stripes Forever," which was a secret code that there was an emergency. The fire raced at incredible speed up the walls and onto the roof.

The Circus Before the Fire

The first performance by Ringling Brothers and Barnum & Bailey® was held in New York City's Madison Square Garden on March 19, 1919. Its most famous clown, Emmett Kelley (1898–1979), performed with the circus from 1941 until 1956. The Greatest Show on Earth® was in Hartford, and thousands of children were looking forward to seeing the elephants, lions, tigers and leopards in the three-rings under the largest tent in the world.

The roof had been coated in a mixture of paraffin and gasoline, which was an inexpensive waterproofing method for canvas. When the flames hit the roof, giant sheets of fire shot into the air. There was little screaming at first, as confused audience members did not realize there was a fire. Then, cries of "fire" began a stampede for the exits. Total panic gripped the crowd, as pieces of flaming canvas fell, igniting hair and clothes. Scores of men, women and children tried to jump from high grandstand seats, and many children were dropped by parents in the hope they would escape. Flaming paraffin was showering like rain and charring the skin of those below. Witnesses said that many deaths were caused by audience members running blindly, with burning canvas covering their heads. A bottleneck formed at the exits, and shrieking patrons trampled smaller children underfoot. All semblance of order was gone.

Within ten minutes, the entire tent roof and supporting poles collapsed, trapping hundreds under the burning canvas. Many victims had become trapped trying to escape over the steel-barred runs through which the animals were led to the ring. They lay two and three deep, and the fire had eaten at their clothes, bodies and faces. One woman who had escaped wrestled with roustabouts in a desperate struggle to re-enter the tent in search of her daughter. "My God, my kid's in there," she screamed. It was to be a common cry from parents who had become separated from their children in the stampede.

Firemen were too late to control the fire and instead focused on hosing down the ruins and burning heaps of bodies. Some people at the bottom of the heaps survived because they were shielded from the flames by the charred bodies above them.

Ambulances, trucks, jeeps and cars raced the dead and injured to the city's four hospitals. The Hartford Red Cross called 1,500 volunteer workers and within an hour teams of nurses and staff assisted at the hospitals and at the Armory, which had been turned into a temporary morgue. Grieving relatives walked through the rows of the dead in an effort to identify a child or a wife. Hearses moved the claimed dead from the morgue to funeral homes over the following days.

Nine days after the fire, the Ringling Brothers Barnum & Bailey circus pulled out of Hartford. They did not return until 1974, 30 years after the tragedy.

OPPOSITE *Partially collapsed animal rings stand at the center of the wreckage, with charred bleachers surrounding them. The cause of the fire was never determined.*

An Unidentified Girl

Many funerals were held on July 8, and flags flew at half-mast. Special masses and memorial services were held on Sunday, July 10. One little girl was never identified and became known as "Little Miss 1565," her body identification number. Various theories abounded as to why she was never claimed, such as her mother died in the fire, and at the same moment her father died in a WWII battle. She was buried along with the other five unidentified dead.

"I would say that it was less than forty-five seconds from the time the first sign of fire appeared until the top of the tent had been consumed."

—survivor

HISTORICAL ACCOUNT

Historian Stewart O'Nan details the attempts to extinguish the fire "Engine Company 2 hooked a double-gated connection onto the hydrant across from McGovern's so the other units could use it, then laid in a thousand feet of line, snaking their way through the crowd, knocking into people, weaving around clumps of bystanders. Engine Company 4 arrived right behind them...On the south side of the lot, the fire from the grandstands was so hot that the light crew had to take cover. The wind was like a blowtorch. Stakes in the ground were burning, guyropes catching the dry grass. Wagons forty feet from the tent caught fire—two of them diesel plants."

The Investigation

Five officials of the circus were arrested the day after the fire and charged with manslaughter, including James A. Healey, Vice-President of Ringling Brothers and Barnum & Bailey, and George Smith, General Manager. They were sentenced to two years in prison for fire code violations. Hundreds of lawsuits were filed, and police prevented the circus from leaving until a judge ruled that the circus should resume performances around the country in order to pay off claims of $4,000,000, which it did in six years.

Empire State Building, New York City

DATE · JULY 28, 1945

DAMAGE · 13 DEAD, 26 INJURED, $500,000 DAMAGE

CAUSE · B-25 BOMBER CRASH

On a foggy Saturday morning in July, Lieutenant Colonel William F. Smith, Jr., 27, of Watertown, Massachusetts, a West Point graduate and veteran of two years combat, was piloting a B-25 bomber. He was on a return trip from Bedford, Massachusetts, with Staff Sergeant Christopher S. Domitrovich, 31, and a Navy enlisted man, Albert Perna. The control tower at La Guardia Airport reported to Smith that the cloud ceiling was very low, with forward visibility at about two miles.

Dropping down out the overcast sky to get his bearings, and possibly mistaking the East River for the Hudson River, Smith suddenly found himself among the skyscrapers of Manhattan. People in office buildings reported that they saw the B-25 winging south. The *Old John Feather Merchant* bomber thundered past Rockefeller Center and barely missed a 60–story building at Fifth Avenue and 42nd Street. It continued south-

The Building Before the Fire

The most famous skyscraper in the world is situated on Fifth Avenue and 34th Street, in Midtown Manhattan. It is 1,250 feet tall, and was designed by Shreve, Lamb and Harmon, in an Art Deco style. Construction was completed in one year, and it opened on May 1, 1931.

bound and one witness told *Newsweek* he "thought the pilot tried to bank away." It was too late. As hundreds of horrified people watched, the plane hit the north side of the Empire State Building at an estimated speed of 225 miles per hour. The ten-ton plane hit between the 78th and 79th floors, ripping an 18-by-20 foot hole in the steel and masonry.

In a flash, the gasoline tanks exploded and flames leaped inside and outside the building. Burning gas flooded the elevator shafts, and there were smaller explosions on the floors below. Part of the fuselage and

"Tower to B-25—we're unable to see the top of the Empire State Building."
—La Guardia Airport control tower

LEFT *The Empire State Building can be seen through this hole in the roof of a neighboring penthouse. The hole was caused by falling debris from the B-25 bomber.*

OPPOSITE *The plane wreckage protrudes from a large hole in the north side of the Empire State Building at the 78th and 79th floors.*

LEFT *Smoke and flame envelop the upper stories of the Empire State Building following the crash of a B-25 Bomber. Smoke also rises from the ground, where parts of the plane fell. Firemen fought the blaze from the floors above, pouring tons of water down on it.*

one motor skidded across the floor, smashed through seven walls and came out on the south side of the building. It hurtled across 33rd Street and through the roof of a penthouse on a twelve-story building. There, it started a $100,000 fire in a sculptor's studio.

The 78th floor was unoccupied, but on the 79th floor women and girl workers were busy at the War Relief Services of the National Catholic Welfare Conference. They never had a chance. Many were burned beyond recognition. The body of a man who worked on that floor was found on a ledge on the 72nd floor—he had apparently been blown out a window.

On the streets 913 feet below, hunks of metal and stone fell as far as five blocks away. A section of a wing was thrown a block east, to Madison Avenue. Gasoline fumes popped in flash explosions four and five floors below. Thick, acrid smoke soon filled the upper floors. Firemen, fighting the city's highest fire from the floors above, poured tons of water on the flames.

Scores of police guarded the roped-off streets, and crowds gathered throughout the day and night to stare up at the building.

By the start of the work-week on Monday, repair-work on the $500,000 damage was well under way. On all except two floors, 15,000 occupants were back at work.

Miraculous Escapes

A woman whose elevator had been dropped to the basement after wreckage sheared the cable was dug out alive. Workers on other floors were trapped behind buckled doors and suffered smoke and burn injuries. There were 26 injured.

La Salle Hotel, Chicago

DATE · JUNE 5, 1946
DAMAGE · 61 DEAD
**CAUSE · ELECTRICAL SHORT
        CIRCUIT**

According to witnesses, the fire broke out about
12:15 a.m. in the Silver Lounge off the north side of the
lobby.

ABOVE *Firemen used a Mack truck, with mounted
aerial ladders and grounded extension ladders, to
rescue guests. Black fire hoses can be seen reaching
from the trucks to various floors.*

The Hotel Before the Fire

The 22-story La Salle Hotel was "the largest,
safest and most modern hotel west of New
York City" when it was built in the heart of
Chicago's business district in 1909.

It had a sprinkler system, chemical fire
extinguishers on each floor and the staff had
regular fire drills.

The Very Reverend William J. Gorman, Chicago Fire Department Chaplain, gave this account to *The New York Times:* "When I came into the lobby it was a roaring inferno. Five persons were found dead by us between the eighteenth and twenty-first floors. Smoke probably killed them, because they weren't burned. I found one woman clutching a child in her arms and lying dead on the roof of the lobby."

The Heroes

Robert Might, a sailor, rescued at least twenty injured people from the upper floors, by dragging them down the fire escapes. Hundreds of frightened guests were saved by firemen who climbed up ladders from the lower floors. Julia Berry, a 44-year old hotel switchboard operator, although surrounded by smoke, calmly called room after room to notify guests of the fire. She was found dead at her post by firemen, and was later given a posthumous award for bravery.

A Miraculous Escape

Anita Blair, a 23-year-old blind girl, smelled the fire outside her 11th floor room, put on her robe, and was led by her seeing-eye-dog, Fawn, through the hysterical crowd to a fire escape. "We followed the crowd around the corner and then a man helped me and Fawn over the sill to the fire escape. Fawn had been nervous and partially blinded by smoke, but once on the landing, he calmed down completely. No one crowded or shoved me."

"Every precaution had been taken to insure the safety of our guests and the public."

—Roy Steffen, President of the La Salle Hotel

After consuming the rugs, couches, draperies and walnut paneling of the lobby, the flames shot up the elevator shaftways, pulled up by drafts from open windows and doors, killing guests in their rooms. Most of the victims on the upper floors were suffocated by dense smoke, which filled the stairways and hallways. Some guests fled through the corridors toward fire escapes, which became blocked in the mass struggle. Many guests trapped on the upper floors leaped in terror, to die on the street or hit lower ledges.

Eugene Freemon, Chief of the First Fire Battalion, was killed in the collapse of the mezzanine floor. Other noted dead were Edward J. Schneidmann, Mayor of Quincy, Illinois and Perry W. Swern, one of Chicago's best-known architects.

The blaze was declared under control at 4:07 a.m., but firemen were still battling smoldering embers that afternoon. More than 1,000 people were rescued, which indicated that many of the fire precautions had worked.

OPPOSITE *Guests at the La Salle come down a fire escape. Following the Iroquois Theater fire in 1903, Chicago building codes required fire escapes on all public buildings, including hotels.*

The Investigation

Investigations afterward placed the cause of the fire as crossed wires behind a false ceiling in the Silver Lounge. It also concurred with the testimony that the varnish and lacquers on the wood paneling contributed most to the deadly smoke, which suffocated many of the victims. In the aftermath of the fire, two Loop theaters and five clubs were closed for fire code violations.

LEFT *Firemen use aerial ladders to rescue people from the worst hotel fire in American history. Ladders are limited in height by the pressure caused by weight at the tip, which can bend or break it. These ladders did not reach the upper floors, and dozens of guests leaped to their death.*

OPPOSITE *Firemen and rescue workers carry one of the 119 dead out of the hotel. The Winecoff had been advertised as being "fireproof," and so did not have any fire escapes or a sprinkler system, which are mandated in today's buildings.*

Winecoff Hotel, Atlanta

DATE · DECEMBER 7, 1946
DAMAGE · 119 DEAD, MISSING, 89 INJURED
CAUSE · UNKNOWN

The worst hotel fire in US history was discovered at about 3:30 a.m., by an elevator operator who smelled smoke coming from the fifth floor and gave the alarm.

The 285 guests in its 194 rooms woke to screams of "Fire!" The elevator shafts and open stairways acted as giant chimneys, drawing the flames up and spreading them through interior hallways. Trapped guests were seen screaming at the windows for help. Several guests crawled out on ledges, waiting for firemen to rescue them. Some were seen lowering themselves from one floor to another on tied together sheets. One man was observed trying to reach a fireman's ladder when two people who had jumped hit him from above and all three plunged to their deaths.

Thousand of spectators jammed the area, and occasional shrieks from the crowd could be heard when someone leaped to avoid the flames. A woman in a flaming nightgown jumped from an upper window, but landed on overhead wires. There she hung, in flames, until finally her body toppled to the ground. Another woman on the seventh floor threw her two small children to their deaths and then jumped to her own. At one time, a half-dozen bodies lay on Peachtree Street,

"It was worse than anything over there…you're just helpless in a hotel room with roaring flames all around you."
—Maj. General P. W. Baade, WWII 35th Div., Winecoff fire survivor

LEFT *Because the ladders could only reach the ninth floor of the fifteen-floor building many people jumped to escape the flames. This woman's fall was broken by the marquee structure and she was one of the lucky survivors. This photograph won a Pulitzer Prize.*

The Victims

Bodies were sprawled in the hallways, and in many rooms men and women had been asphyxiated. A reporter found a woman kneeling in the bathroom, with her three children gathered into her arms, frozen in time by the flames, like a blackened statuary. Ankle-deep water swirled through the hallways, as firemen brought down body after body. In one room, four teenage girls lay as though sleeping; in another, a woman lay peacefully with her head on a windowsill, with her purse in her hand; in another, an abandoned canary sang; in one room, an uncooked turkey wasn't even singed. On the steamy tenth floor, firemen found the body of W. F. Winecoff, 70, who built the hotel in 1913. He had lived there since his retirement in 1934.

opposite the theater where the premiere of *Gone with the Wind* had been staged.

The flames were finally brought under control, after six hours of effort. Firemen began a grim room-by-room search. Many rooms were a mass of charred wreckage.

By mid-morning hospitals were jammed with the injured, and the dead had been taken to morgues to be identified. Grieving relatives searched the morgues for a ring, scar or birthmark to identify their loved one.

Shocked and horrified residents jammed the area to view the gutted hotel. The sheets and curtains that had been used by guests in their desperate efforts to get to safety still hung from the empty windows, and fluttered in the breeze. Praise was heard for the valiant firefighters, as well as the hospitals and volunteer organizations.

Texas City, Texas

DATE · APRIL 16–18, 1947
DAMAGE · 500 DEAD,
$50,000,000
DAMAGE
CAUSE · UNKNOWN

At 8 a.m. a small fire was discovered aboard the French freighter *Grandcamp*, which had docked at Texas City five days earlier. At 8:30 the crew of 41 abandoned ship, and fire alarms were sounded. The crews of nearby ships, dockworkers, laborers, schoolchildren and at least 100 townspeople thronged to the docks to watch as twelve fire trucks arrived and attempted to quench the fire.

The ship had been loaded with bags of ammonium nitrate the day before. The white, sugary chemical is used as fertilizer, but can also be used in the making of TNT. When the fire reached the ammonium nitrate, the

BELOW *Coast Guard cutter Cris fights the flames on the burning docks.*

The City Before the Fire

This industrial port town of 15,000 people, ten miles from Galveston, is home to the chemical company Monsanto. In 1947 it produced styrene, a chemical used in manufacturing synthetic rubber and plastic. The town was also home to the Pan-American Refining Company, Republic Oil and Continental Oil, which had numerous oil storage tanks.

frightful explosion literally blew the ship to bits. Hunks of steel and wood slashed at people on the streets, and the concussion killed many nearby spectators. Out on the bay, passengers aboard the Galveston ferry saw a spout of water leap 4,000 feet high into the air. Wind whipped the flames, and the fire spread inland.

One dockworker told *Newsweek*, "A big oil barge which had been tied up at the dock flew through the air and landed nearby." A warehouse worker said, "There was a second explosion and the roof and walls of the warehouse were coming down around me...one man with a leg blown off was screaming in pain."

"We plan to build the most modern port on the Gulf of Mexico."
—Walter Sandberg, President, Texas City Terminal Co., following the fire

Storefronts in the downtown business district collapsed and nearby grain elevators were set ablaze. Oil tanks were ignited by flaming splinters, and the air was full of glass, girders, boats, barges and fragments of human beings.

Texas City's fire department had been wiped out in the first blast and soon 2 miles of the waterfront were on fire. The Monsanto plant caught fire, sending gases from petrochemical storage tanks into the air, suffocating fleeing residents. By noon, a gigantic black cloud

The Cause of the Explosion

Ammonium nitrate is easily made by adding ammonia to dilute nitric acid. Chemically, ammonium nitrate is a salt, a combination of a base with one nitrogen atom and four hydrogen atoms, and an acid with one nitrogen atom and three oxygen atoms. These are kept in a fragile balance until a disturbance such as heat or shock shatters the barrier. Then the oxygen atoms grab hydrogen atoms and nitrogen atoms grab oxygen atoms. This process generates enormous heat and the salt flashes into gases: nitrous oxide (N_2O) and superheated steam (H_2O).

hovered over the city. Fire trucks, ambulances and the National Guard arrived from Galveston, Houston, Port Arthur and Beaumont to assist. Attempts to quench the fire were useless and throughout the night thousands of dazed victims, whose homes had been destroyed, plodded through the streets, picking their way past bodies.

At 1:10 a.m. the freighter *High Flyer*, which was moored near the *Grandcamp* and also loaded with ammonium nitrate and sulfur, exploded. The *Grandcamp*, after burning all day also blew up. The *High Flyer* then drifted against the *Wilson B. Keenes*, causing it to detonate. Fifty men fighting the fire were instantly killed. New explosions of nitrate, sulfur and chlorine from the various storage tanks of the chemical companies along the waterfront lit up the sky. Fires continued to burn in the oil tanks all day Thursday, and late that night a butane gas tank exploded, feeding the fire.

By Friday morning, four days after the start of the fire, Deputy Mayor John Hill was able to announce that, because of the valiant efforts of the firefighters, "All fires are under control." Galveston and Houston hospitals couldn't accommodate all the injured, and in the Houston municipal auditorium rows of wounded lay on the floor.

Over 150 embalmers, from as far as Fort Worth, worked in foot-deep blood in the former McGar's

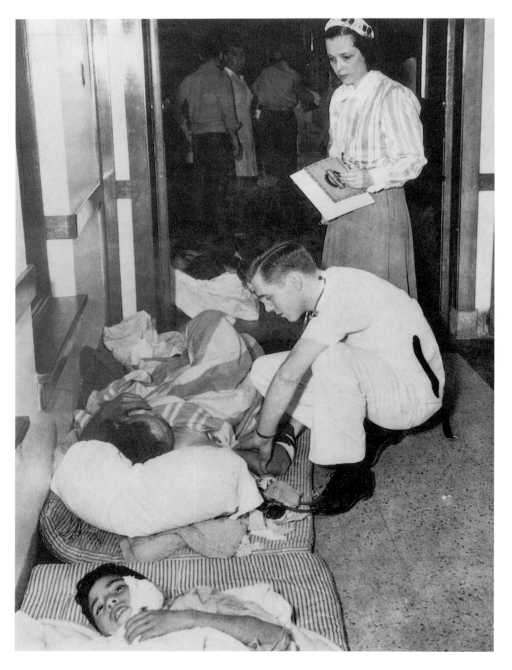

LEFT *An intern at a Galveston hospital aids an injured victim who was placed in the front hallway of the overcrowded facility.*

ABOVE *Smoke billows from the Mousauts Chemical plant.*

Garage, attaching tags to the left ankle of each victim. The tags read, "You have violated a traffic law. Port Arthur Police," because they were the only ones available. Trucks carrying bodies stacked five and six deep arrived all day. The high school gym was turned into a morgue and relatives bent over rows of bodies to identify a husband, wife, son or daughter.

On June 22, 1947, in donated coffins, the bodies of 63 unidentified dead were buried in memorial cemetery in numbered graves.

A million dollar relief fund was set up to pay for counselors, medical care and retraining. In addition, almost four million dollars in insurance payments to property owners assisted in the rebuilding of Texas City.

One year later, the Monsanto chemical plant was being rebuilt and enlarged, and shattered lives were returning to a semblance of order. Approximately 700 houses had been repaired, 500 new homes were built. The blast also led to improvements in disaster preparedness including communication and safety procedures. Changes in technology, such as improved foams, have enhanced the firefighter of today's chance against large and dangerous chemical fires.

Restoring Order

Bulldozers, tractors and crews of laborers began to clear rubble. Miraculously, some people were found still alive in the wreckage of the Monsanto plant. Volunteer organizations such as the Salvation Army and the Red Cross supplied medicine, food, housing and other relief. The oil companies brought in men and equipment to help with the debris. One survivor told *Newsweek*, "Some people will leave town and go back to the farm... The rest of us will build the town back."

St. Anthony's Hospital, Effingham, Illinois

DATE · APRIL 5, 1949
DAMAGE · 75 DEAD
CAUSE · UNKNOWN

"Newborn babies among victims as midnight blaze roars through institution swiftly."

— *The New York Times* headline,
April 6, 1949

The nation's second worst hospital fire began at 11:50 p.m. at St. Anthony's Hospital, and quickly became an inferno, trapping in their beds many patients with their limbs in casts or suspension slings. The Illinois Fire Marshall Pat Kelly said that the origin of the fire was undetermined. He also said that a combination of cellulose wall material, old and fresh paint and unprotected wall openings helped the rapid spread of the flames.

LEFT *Heavy smoke pours out of St. Anthony's Hospital, while firemen use ladders to rescue patients, and fire hoses snake on the ground.*

The fire department reached the scene within fifteen minutes, but according to Assistant Chief Charles Chamberlain in *The New York Times,* "We could see the building was gone, so we just tried to save people." Nuns gathered mattresses to break the fall of about a dozen who leaped from the windows of the three-story brick building.

FIRE-FIGHTING TECHNOLOGY OF THE TIME

Hospital fires can be very dangerous, due to the chemicals and gases kept there. Today, firefighters would don hazardous material protective clothing and breathing apparatus before entering a hospital fire. At mid-20th century, these precautions were not yet common.

LEFT *Firemen search the ruins for bodies. Investigations showed that many victims suffocated due to the poisonous smoke caused by new and old paint. Note that the walls have burned down to the brick.*

For more than an hour the night was filled with the screams of bedridden patients. Many reached windows, stood for a moment silhouetted against the flames and fell back into the fire. Some perished as they crawled through smoke- and flame-filled corridors.

Mrs. Wenton Sidner was in the delivery room awaiting the birth of her baby. She jumped screaming from a second-floor window and broke her back and right arm. The baby was born dead an hour later. Mrs. Arnold Adderman was luckier—she was also in the delivery room, but she kicked out a screen, crawled onto a porch roof and down a ladder. An hour later she gave birth to a boy. Ten newborns and three other babies perished in the fire.

The fire was extinguished within a few hours, but firemen suspended digging for victims on the evening of the 6th because of the building's shaky walls. On the 7th the final victim was taken out. Among the dead were seven hospital employees, eight nurses, four nuns and the chaplain.

Governor Adlai Stevenson inspected the ruins on the 6th and said that the hospital had complied with all state fire requirements. A representative of the state fire marshal had inspected the building about six weeks prior to the fire.

The grief-stricken town began the funerals on the 7th, and on April 12 a solemn memorial to the dead was held. Mayor H. B. Rinehart read a letter of condolence from President Truman, who also pledged over $2,000,000 for a new hospital.

The Reaction

The fire prompted calls for more fireproof materials in hospitals, metal furniture and non-combustible laboratory equipment.

Katie Jane Memorial Home for the Aged, Warrenton, Missouri

DATE · FEBRUARY 17, 1957
DAMAGE · 72 DEAD
CAUSE · UNKNOWN

Mr. W. S. O'Sullivan operated the Katie Jane Memorial Home in a 2–story brick and wood frame building with no fire alarm or fire escapes. There were 155 patients ranging in age from 50–90.

The flash fire began about 3:45 p.m. and sped through the building with such speed that it was all over within 15 minutes. This led to speculation that the cause might be arson, but this was never proven.

Then an explosion, probably from gas lines catching fire, sent flames as high as sixty feet, and smoke could be seen thirty

BELOW *The roof collapsed, taking some of the masonry walls. Building failures can result from deteriorating mortar heated to extreme temperatures and collapsing.*

Mrs. Knigge, the wife of the coroner, told the Associated Press, "I heard screams and ran from my home to the building…by the time I got there it was all over."

> "I've spent $30,000 trying to fix it up to avoid something like this."
>
> —W. S. O'Sullivan, operator

miles away. "It was just an inferno—like taking four brick walls and an oven and building a fire as big as you can," said John Berkholtz, a wholesaler who happened to go by when the fire started.

It was hours before the search for bodies could begin because of the continued intense heat from the ruins. To get to the bodies, firemen pulled down the walls as the heat began to subside. Emergency supplies were rushed from St. Louis to Washington Hospital in Warrenton.

LEFT *Rescuers probed the smoldering ruins for victims. Note the charred steel beds in the foreground. They were standard "fireproof" items for hospitals and nursing homes, but the bedding burned.*

The Investigation

The following day it was revealed that the home had been operating without a license, although the prior week it had been inspected and approved. The police investigation revealed that a nursing home operated by O'Sullivan's sister had burned in 1952, five years earlier, killing eighteen patients in Hillsboro, MO. However, W. S. O'Sullivan was not indicted.

By February 20, all the bodies had been removed from the wreckage. Efforts to identify the charred remains at the morgue were being made by relatives and friends. The coroner's report of March 4th found that the fire was of unknown origin and no charges were pending.

Our Lady of the Angels School, Chicago

DATE · DECEMBER 1, 1958
DAMAGE · 90 CHILDREN, 3 NUNS DEAD
CAUSE · UNKNOWN

"We pray for comfort from heaven for those in sorrow."

—Cable from Pope John XXIII

The fire was discovered at about 2:30 p.m., just before the end of the school day. Later investigations showed that it started in a trash barrel at the foot of the northeast stairwell, possibly from a discarded cigarette. It spread rapidly up the stairwell and into the airspace between the ceiling and the roof directly over the students' heads. For a number of minutes, nuns, lay-teachers, and students were unaware there was a fire. Suddenly, smoke swirled into the classrooms and flames appeared through the ceiling. Panicked children began to open windows and leap before firefighters arrived.

The Chicago central fire alarm office received the first call at 2:42, and immediately seven fire companies were

The School Before the Fire

The school and adjoining church and rectory had been built in 1939. The two-and-half story school was shaped like a U, with the north wing fronting on North Avers Avenue and the south wing on West Iowa Street, connected by an addition of 1951 called the Annex. The school contained 24 classrooms, and was built with wood framing covered with plaster, under a brick exterior. It had acoustical tile ceilings and wood trim throughout. There were open stairways, no sprinklers and only one fire escape on the Annex. Additionally, the fire alarm did not connect to the fire department, but only rang at the school.

dispatched, but despite the fact that they got there in five minutes, the fire had already been raging for a half hour.

Firemen raised ladders and began pulling children out the windows and off the sills. The second floor windows were jammed so tightly with children that the children in front were crushed while those behind struggled to avoid the flames. According to David Cowan and John Kuenster, historians, "When [fireman Charlie Kamin] reached the window he came face to face with the most terrifying sight in his life. Bunched before him in the smoke was a pile of eight-graders stacked in more layers than he could count. Dozens of hysterical kids were crowding the opening, all screaming and pulling at each other in a mad fit to reach safety. Burning debris was falling from the ceiling, and those nearest the ledge were fighting to grasp hold of the ladder…Kamin turned back to the window, and from that point on, worked like a robot, reaching inside

LEFT *Firefighter Richard Scheidt, his face drawn in anguish, carries the limp, dead body of John Jajkowski, age 10, out of the school. Firefighters sobbed uncontrollably at the horror, but persisted in their mission.*

RIGHT *The body of a girl is carried down a ladder by a fireman. Many children were saved by heroic firemen, but 90 died.*

the room and grabbing students…the room had finally reached its flash point. The air was igniting. And then it blew…He then saw the rest of the children in the window just wilt and turn dead, like a bunch of burning papers."

At about 3 p.m. portions of the roof caved in, trapping children in their classrooms. Hysterical mothers and fathers pleaded to be permitted to enter the building. Nuns and priests, including Monsignor Joseph Cussen, in charge of the parish, assisted in treating the injured.

In order to pick up the injured, firemen had to cut down trees and bushes to allow ambulances to approach the school, because the streets were clogged with emergency vehicles and the cars of frantic parents.

By 3:30 firemen succeeded in extinguishing the flames by slowly moving into the school, and by 3:45 the search for the dead began. Firemen found toppled desks and heaps of burned children, some almost incinerated. Cowan and Kuenster give the counts: "Room 212, Grade 5, 28 dead, 21 injured; Room 210, Grade 4, 30 dead, 15 injured; Room 208, Grade 7, 10 dead, 13 injured; Room 211, Grade 8, 25 dead, 17 injured; Room 209, Grade 8, 2 dead, 8 injured; Room 207, Grades 5 & 6, 0 dead, 1 injured."

The front page of the *Chicago American* newspaper read, "Chicago Mourns."

The effect on the families of those who died, as well as those who survived, was devastating. Some people never spoke about it, while others relived the fire for years. Some grieved privately and others blamed the nuns for not getting the children out. The close community was shattered. Heroic firemen who had saved hundreds of children could remember the faces of the children years later.

The Investigation

The inquest afterwards did not produce a definitive cause for the fire. There was some indication that it was a cigarette, but the cause is still listed as "unknown." A few years later, a teenaged firebug signed a confession, but the juvenile court did not find cause to believe him. Still later, another man confessed, but this was also discounted. Many investigators still believe the fire was intentionally set by some unknown person.

The Victims

At about 4:30, Mayor Richard Daley and Archbishop Albert Meyer arrived to inspect the ruins. Rows of sheet-covered bodies were laid on the ground awaiting transport to the morgue. Parents continued to search for missing children until the last victim was removed at 7 p.m. St. Anne's Hospital, where most of the injured children were taken, responded quickly and efficiently. Triage was done and the worst burned were immediately treated. According to *The New York Times* on December 3, "There were offers of blood, of money, of skin to cover the scars."

Fire Behavior

A flashover occurs when trapped smoke and fire collecting at the ceiling level radiates downward, heating the contents of the room. The overall temperature increases, and in a very short time, usually seconds, the entire room and all its contents spontaneously ignite from the intense heat.

BELOW *A fireman probes the debris in one of the charred classrooms on the second floor. Wooden desks, floors and walls contributed to the blaze, which collapsed the roof (on right), trapping many students.*

"The people of the nation as well as the City of Los Angeles, feel a deep sense of relief as order is being restored to the frightened streets of that city."

—President Lyndon B. Johnson

LEFT *Sunday morning after the riots reveals the devastation of the looting and the fires. Portions of Watts have never recovered.*

Watts Riots, Los Angeles

DATE · AUGUST 11-16, 1965
DAMAGE · 34 DEAD; $200 MILLION
CAUSE · ARSON

Hundreds of fires were set by rioters in Watts. One raged through a 3–block area while 14 fires were raging on 103rd Street alone. The 150–block area of Watts looked like a war zone, with streets littered with burning debris, demolished storefronts and a heavy pall of

The Beginning of the Riot

The rioting began after a routine drunk driving arrest of an African American by white officers. African-Americans at the scene complained of police brutality and began throwing rocks. Chief of Police William H. Parker and Mayor Samuel Yorty requested the aid of the National Guard. On Saturday, August 14, 2,000 heavily armed National Guardsmen entered Los Angeles, battling with fixed bayonets against rioters, and at least once using machine guns. On the fourth day of rioting Governor Edmund Brown, with the cooperation of President Lyndon B. Johnson, sent in 20,000 National Guardsmen and vowed to restore order. Troops moved block-by-block, removing everyone from the streets after the 8 p.m. curfew. Crowds numbering in the thousands chanted and cursed the Guardsmen.

LEFT *Three stores burn to the ground on Avalon Blvd. Hundreds of similar fires raged uncontrolled until the National Guard was called in and restored order.*

smoke hanging over the city. Overturned and burning cars blocked streets, and smoke rose into the air from fires set by looters. Most fires roared uncontrolled because rioters were throwing Molotov cocktails at the firetrucks and snipers were shooting at firemen.

A fire captain told UPI, "We don't have the manpower to extinguish them." There were over 200 fires set on the 13th and over 350 fires begun on the 14th. "All the big stores have been destroyed or damaged," the Fire Chief reported. One of the biggest fires leveled the Friendly Furniture Company factory and quickly spread to neighboring buildings. One man said, "These people are burning up their jobs and homes."

By the 16th, after three days of rioting, the crowds had been dispersed, order had been restored and the fires either burned themselves out or were extinguished by exhausted firemen.

Firefighters Under Fire

Firefighter safety is dependent on many factors. Half the injuries and deaths occur as a result of stress and panic at fire scenes. In the 1960s firefighters were unprepared for assaults. Now firefighters wear protective clothing, such as bulletproof vests and flak jackets, when called to a potentially dangerous scene.

23rd Street Fire, New York City

DATE · OCTOBER 17, 1966
DAMAGE · 12 FIREMEN DEAD
CAUSE · UNKNOWN

The worst single loss of life in the history of the New York City Fire Department happened a week after "Fire Prevention Week."

Upstairs in a building on 22nd Street, at 9:36 a.m., Mrs. Brown smelled smoke and saw fire coming from a 2–story extension built between her building and 6 East

---

OPPOSITE *Firemen pour immense streams of water on buildings during the fatal five-alarm fire. Illegal walls in the cellar contributed to the ferocity of the fire, causing a floor collapse, which killed 12 firemen.*

BELOW *Emotional lines of firemen bow their heads and place their helmets over their hearts as one of their comrades is carried from the destroyed building. It was the worst loss of life for the Fire Department in its history.*

23rd Street, a brownstone containing a Wonder Drugs store, Fan's Lingerie Shop and Barton's Candy Store. She reported the flames, then escaped with her husband and children.

Minutes later Engine Companies 3, 14 and 16, along with Ladder Companies 3 and 12, rushed to the scene, hoping to catch the fire before it spread through the building. Battalion 6 and Deputy Chief Thomas Reilly, Ladder Company 7 and Engine Company 18 also arrived in support of the effort.

The building extension emitting the smoke was surrounded on all four sides by taller buildings, and therefore could not be properly inspected from the outside, substantially increasing the danger to the firefighters. Firemen were sent to the basement of Wonder Drugs to find the fire and identify, if possible, what was burning.

Because of an illegally built cinder block wall, constructed in the cellar about three quarters of the way down the length of the upstairs store, the firefighters

"The 12 Never Had a Chance—We All Died a Little in There"
—*New York Daily News* headline

Fire Behavior

In commercial building fires, there is the added danger of chemical fires, as well as dangerous concentrations of stored flammables such as paper products, boxes and plastic wrap, so it is extremely important for firefighters to be cautious entering these dark and smoky conditions to find the source of the fire.

didn't see the fire. This gave the fire, which was located on the other side of the wall, in the back quarter section of the cellar, time to build to extreme heat. Because the fire was in the basement, with relatively little ventilation, the heat was trapped and was rapidly increasing, a situation that firefighters refer to as a firebox. Cellar fires are very difficult to fight because firefighters have to approach them from above, directly against the rising heat. The key is to ventilate as quickly as possible, releasing the heat and smoke.

Twelve firemen raced to the back-third section of the drugstore, still trying to find the fire, when the flooring suddenly collapsed. Tragically, ten firefighters fell to their deaths in the inferno raging in the cellar below. Two others were instantly killed at the very edge of the collapse, probably by the flames and heat, which were fueled by the new influx of oxygen shooting up through the hole in the floor.

Mayor John Lindsay arrived to find heartsick firemen attempting to rescue their brethren. The bodies of the two firemen were removed from the first floor precipice around 1:30 a.m. It turned out that the brownstones on East 23rd and 22nd Streets were illegally connected by a previous owner, to create apartments on the second floor. The cellar had also been illegally divided by two cinder block walls, creating a maze-like setup that caused the fatal delay in the firefighters' search efforts.

The following morning, October 18, a crowd of several thousand firefighters and ordinary citizens watched in grim silence as the bodies of the remaining ten firemen were removed from the pit by members of their respective units.

On October 21, 1966, Fifth Avenue was a "sea of blue," as 20,000 firefighters from all over the world came to pay their last respects. Thousands of onlookers lined the streets as the cortege carrying the bodies stopped first at St. Thomas Episcopal Church and then again at St. Patrick's Cathedral.

In 1976 Ralph Feldman, of Engine Company 37, fashioned a monumental sculpture, with burnt wooden beams and charred bricks held on the cross beam by a skeletal hand of steel. This "Memorial to Twelve" was dedicated during Fire Prevention Week and is on permanent display at the Cathedral of St. John the Divine.

Close Calls in a Dangerous Situation

The rescue team, led by Jack Donovan, fought towards the fire. Donovan, whose vision was obscured by the smoke and steam rising from the hoses, reached the fire and fell into the pit. He managed to cling to his fire hose, dangling over the flames as he was pulled to safety by his backup team. Engine Company 16 was fighting the fire on the roof and was trapped by flames. They narrowly escaped when a ladder truck rescued them just before the roof collapsed.

Structural Failure

At 800 degrees, the steel supports of buildings start to expand and buckle, and at 1,100 degrees they fail.

LEFT *An injured firefighter is tended to by fellow firefighters in a scenario that was repeated many times during the blaze.*

ABOVE *Portrait of the Apollo 1 astronauts standing in front of a Saturn 1 booster; (left to right) Virgil (Gus) Grissom, Edward White and Roger Chaffee.*

Apollo 1

DATE · JANUARY 27, 1967
DAMAGE · 3 DEAD
CAUSE · SPARK OF UNKNOWN
ORIGIN IN ENVIRONMENT
OF PURE OXYGEN

"Three valiant young men have given their lives in the nation's service. We mourn this great loss and our hearts go out to their families."

—President Lyndon B. Johnson expressing the sorrow of the nation

The three-man crew of Apollo 1, Virgil I. Grissom, Edward H. White and Roger B. Chaffee were killed in a flash fire when they were trapped in a test capsule on the launch pad.

The astronauts boarded the spacecraft on top of a Saturn 1–B rocket some five and a half hours before the fire. They were participating in a full-scale simulation of the flight scheduled for February 14. The simulated countdown had reached T-minus ten minutes

The Space Program Before the Fire

The Apollo Space program was announced by the U.S. National Aeronautics and Space Administration in May, 1961. Its aim was to land an American on the moon by 1970, which was finally achieved by Apollo 11 in July, 1969, when astronaut Neil Armstrong became the first man to set foot on the moon. He declared, "One small step for man, one giant leap for mankind."

when the fire broke out. The astronauts were trapped behind sealed hatches in the super-oxygenated atmosphere. High concentrations of oxygen can cause some materials to ignite spontaneously. It allows material to burn that would not burn under normal circumstances. In addition, 100-percent-pure oxygen was being fed into the spacesuits, so there was fire both in the air and in the spacesuits. The last words of the astronauts were, "Fire in the spacecraft," and then they died almost instantly.

The inside of the capsule was reported to be a mass of flames. It took ground crews monitoring the test about four minutes to rush up the high-speed elevator to try to rescue the crew. Despite dense smoke, the ground crew managed to open the hatch, but it was too late. The cockpit interior was described as 75 percent destroyed. Twenty-seven workmen suffered from smoke inhalation in the rescue attempt.

The accident caused NASA to deploy several unmanned flights while new precautions and procedures were implemented. The first manned flight after the accident was Apollo 7, launched on October 11, 1968.

ABOVE *Exterior of the charred Apollo 1 space capsule. The astronauts died in the command module as a result of an electrical spark in the oxygen-rich atmosphere.*

The Investigation

The 7–man investigation board focused on the cause of the fire. A spark igniting the pure oxygen in the capsule was viewed as the probable cause, although how a spark was introduced into the capsule was a matter of conjecture. It could have been electrical or it might have been a dropped object.

The Store Before the Fire

The five-story building on the Rue Neuve was built in 1897. It had a central atrium surrounded by balconies. The store had been promoting a "U.S. Parade" of merchandise, ranging from bluejeans to paper dresses. The promotion had been the target of protests by the pro-Peking Communist group "Action for the Peace and Independence of Peoples." Pickets and calls for a boycott of the store took place the week before the fire.

L'Innovation Department Store, Brussels, Belgium

DATE · MAY 22, 1967
DAMAGE · 322 DEAD
CAUSE · UNKNOWN

Brussels' second largest department store, L'Innovation, was filled with approximately 1,000 shoppers and employees when a suspicious fire completely destroyed it.

The fire in L'Innovation was discovered around 1:30 p.m. Some witnesses said it started in the third-floor camping department, others swore it started in the fourth-floor restaurant, while other reports stated it began in the fourth-floor furniture department. Store management issued a statement saying it had started in a small closet in the first-floor children's wear department.

The alarm system malfunctioned, and many people did not know there was a fire until they saw the smoke and flames. Cries of "fire" created a panic. Shoppers on the first floor rushed out the door, but the fire spread

Fire Behavior

Atrium spaces are very dangerous if a fire starts at the bottom. The large amount of oxygen available to fuel the fire allows the fire to grow rapidly. A chimney effect pulls the smoke and flames toward the ceiling. With no ventilation in the roof, the fire circles back and enters the open balconies, which typically do not have fire walls.

ABOVE *A man hangs out a window and a woman stand on a ledge, about to leap. The fire alarms failed, and firemen had difficulty reaching the store due to the narrow streets with cars parked on them.*

Heavy black smoke and flames blocked the main stairway. A young man in the fourth-floor restaurant reported to *The New York Times* that it "was suddenly transformed into a gas chamber," and that he did not know how anyone survived; "I saw them fall around me."

up the interior atrium, trapping many shoppers on the upper floors.

Firebrands spread by the wind flew onto the roofs of neighboring buildings and ignited them. Two blocks of older buildings, which had been scheduled for demolition, burned.

The first firetruck arrived within five minutes, but it was a half hour later before other, larger trucks could get through the tiny streets surrounding the store. They were blocked by cars parked half on the street and half on the sidewalk. The Brussels firefighters had only five aerial ladders, 19 fire trucks and 317 men.

Shoppers rushed to the windows and some jumped to their deaths before ladders could reach them. Firemen, with the limited supply of aerial ladders, had to choose which people to save from the dozens who were begging for help.

Deputy Fire Chief Jacques Mesmans told *The Times* of London, "I picked up one poor woman who jumped out of a second floor window and shattered both legs…one man was transformed into a living torch before my eyes as he hesitated to leap from a high window."

A whole side of the store collapsed outward, covering fire engines with blazing rubble, but no firemen were injured.

By the next day the fires were extinguished, and searchers were combing the debris for the dead and missing. King Baudouin of Belgium, Premier Paul Vanden Boeynants and Leon-Joseph Cardinal Suenens, the Primate of Belgium, visited the ruins.

LEFT *Aerial view of smoke billowing from the fire that destroyed one of the largest department stores in Brussels.*

The Investigation

Investigators questioned 30 known left-wing extremists because of reports of bomb threats, but in the end no exact cause was determined.

Riots, Newark, New Jersey & Detroit, Michigan

"Burn, Baby, Burn"
—rioter in Detroit

DATE · JULY 11–27, 1967
DAMAGE · 23 DEAD, $10.2
MILLION IN NEWARK;
43 DEAD, $50
MILLION IN DETROIT
CAUSE · ARSON

The race riots of Newark and Detroit occurred within days of each other and were some of the worst violence of the century. The fires that attended them destroyed major portions of each city, neither of which fully recovered.

Molotov cocktails were thrown into looted stores, which burned uncontrolled. Fire headquarters reported more than 100 fires. Snipers aimed at firemen who continued in their vain attempts to extinguish fires all over the city. Fire Captain Michael Moran was shot off a ladder and killed as he tried to extinguish a blaze on

ABOVE *Police confront rioters before widespread arson devastated the Detroit area.*

Central Avenue. Responding to a pre-dawn blaze on Sunday the 15th on Springfield Avenue, the heart of the ghetto, firemen, again coming under fire, could do little to battle the flames.

Fire Starting Device
A Molotov cocktail (after Vyacheslav M. Molotov [1890–1986], Russian statesman) is usually a bottle filled with gasoline, wrapped in a saturated rag, ignited and hurled into a building or at a target to start a fire. It is classed as an incendiary device and is extremely explosive.

By the 16th the riots in Newark were finished but they were about to break out in Detroit with much more ferocity.

The mood of the crowd grew uglier when rumors spread that police had bayoneted a man and by 1 p.m. rioters were stoning the police and setting fires to buildings. At 4:30 the fire department put out a "3–777" alarm, which hadn't been used since WWII, that meant the immediate recall to duty of all firefighters. Within hours 95% of all firemen were at their stations. There was little they could do on Twelfth St. because they were being pelted by rocks and bottles and coming under fire by snipers. They withdrew behind police lines and left their hoses in the streets.

"It's embarrassing to firemen to see fires going without being able to put them out" said Chief Quinlan.

The fires lit by arsonists, originally targeting white-owned businesses, were now burning large areas of the city. A warm wind fanned the flames and nearly a mile

BELOW *Firefighters continue to battle the fires late July 14th in Newark started during rioting the night before and that morning.*

The Beginning of the Riots

The incident that touched off the Newark riot on Wednesday, July 12, was the injury sustained by an African-American cab driver. *The New York Times* reported, "The violence was believed to have started after the arrest of a Negro cab driver on charges of assaulting a policeman." Witnesses claimed they saw police dragging him into the 4th Precinct Station and rumors spread that he had been beaten. Large hostile crowds of African Americans threw rocks, set fires and looted that night and the following night. By Friday, July 14th, Mayor Hugh Addonizio had asked Governor Richard Hughes to call in the National Guard. Shotgun-toting police guarded firemen who were trying to extinguish a huge blaze at Broad and Market Streets, the center of downtown Newark. Police were instructed by officials to return gunfire. Snipers on rooftops in the Central Ward shot at police while belligerent cops and National Guardsmen used armed tanks and fought pitched battles in the streets. The increasing violence was exacerbated by the fact that the Guardsmen were white, young, scared and totally lacking in riot training.

The Beginning of the Riots in Detroit

At about 4 a.m. on Sunday, July 23rd, Detroit police raided a "blind pig," an after-hours drinking club where a party for several returning African-American Vietnam Veterans was in progress. Eighty-two partiers were arrested and it took over an hour to remove them from the club. A large crowd began to gather and store windows were smashed. Looting began about 8:30 a.m. and continued all day. Police were given no orders and stood by watching as looters, in a "festive mood," ransacked stores. They booed their Congressman, John Conyers, when he pleaded with residents to stay calm. With the Newark riots occurring just a week before, there was a feeling that the authorities had no respect for the African-American community. The attitude held by African Americans that they were not represented in equal proportion to their number increased their anger. Looters defended their acts as a way to deal with "Whitey."

The Reaction

Many questioned why African Americans rioted and burned down their communities. Racial prejudice, poverty, living conditions and other factors were blamed. Historian Ann K. Johnson quotes the results of a survey of prisoners arrested in the riots: "They were young...most had resided in Detroit for many years. Most had jobs...they did not like their living conditions—they didn't like their neighborhoods, threats, the roaches, the filthy tenements and apartments. They wanted equal rights."

of Twelfth St. was a solid sheet of flame. Fires extended down a twenty-block stretch of Grand River.

On the 25th guerrilla warfare broke out as troops struggled to restore order. Sherman tanks and machine guns were being used by troops, who also spread huge bails of barbed wire to block the streets. Huey helicopters swooped overhead with machine gunners stationed at the open doors.

Order was restored by Thursday the 27th and the National Guard was removed on August 1. Detroit was devastated. It was the worst riot in American history up to that time. 43 people were killed and 600 injured. Hundreds of fires had destroyed more than 1,300 buildings. Approximately 5,000 people were left homeless and thousands were out of jobs because their workplaces, hundred of buildings in the heart of Detroit, had been destroyed.

BELOW *A Michigan National Guardsman looks upward for snipers as a firebombed building burns in Detroit. Police and National Guard were deployed to both stop the riots and protect the firefighters.*

A bartender who escaped through a door behind the bar said, "There was panic everywhere. People were running and yelling and the band continued to play."

Club Cinq Sept, Saint-Laurent-du-Pont, France

DATE • NOVEMBER 1, 1970
DAMAGE • 144 DEAD
CAUSE • A CIGARETTE

ABOVE *Local villagers gather to view the ruins of the nightclub, which had no alarm, no telephone and no fire hydrants.*

The small town of Saint-Laurent-du-Pont, 14 miles northwest of Grenoble in southeastern France, was the scene of one of the most heart-rending fires in Europe.

The interior of the one-story concrete-block dance club on the outskirts of town had been lined with highly flammable plastic and papier-maché decorations. There was no telephone to give the alarm and no hydrant or hose connection for fire fighting, probably because the club was outside the city water supply.

The fire, caused by a stray cigarette according to a spokesman for the police, broke out at 1:45 a.m. in a balcony while about 165 dancers were enjoying a new rock group from Paris. The emergency doors had been

"A huge flame leaped into the air and suddenly plunged down to the main floor like a whirlwind. Everybody was screaming, screaming, screaming and suddenly nothing more except the sound of the sirens of the firemen arriving."

—Joelle Dondey, a cashier

padlocked and nailed shut with planks on the outside to keep out any gatecrashers. Two boys ran nearly a mile to give the alarm because there was no phone.

The only door that was open was at the entrance and was blocked by a turnstile. One man who was near the door and escaped with three friends said, "We took a beam and were able to knock a door and pull out a young woman…there were people piled up behind, reaching out with their arms. Five minutes later they were all dead." Another said, "We soaked our jackets in the stream near the dance hall and used them to smother the flames on the people we could pull out."

In five minutes 30 volunteer firemen arrived. They thought at first that most of the dancers had escaped because it was so quiet. Henri Fattalini, a fireman, told the Associated Press, "There wasn't a murmur or a cry. Imagine our horror when the first group succeeded in getting the door open and then felt bodies falling on them."

Breaking into the club, rescue workers found the bodies of 144 victims aged 17 to 25 in charred heaps piled up in front of the turnstile and behind the boarded up emergency exits. Among the dead were two of the three owners of the club and members of the band. Bodies were taken to an emergency morgue at the school gymnasium and grieving relatives walked up and down the rows of coffins searching for their dead loved ones.

ABOVE *The Saint-Laurent-Du-Pont gymnasium filled with coffins holding the remains of the 144 dance hall victims. Grieving relatives filed past to make identification of loved ones.*

The Reaction

Only twenty youths survived this devastating fire. Authorities suspended Mayor Pierre Perrin and a second official. Charges of negligence in safety inspections resulted in increased safety precautions throughout France. As a direct result of the French fire, the British government introduced a bill in Parliament that required fire-fighting equipment in all hotels, dance halls and clubs.

Beverly Hills Supper Club, Southgate, Kentucky

DATE · MAY 18, 1977
DAMAGE · 165 DEAD, 100 INJURED
CAUSE · ELECTRICAL SHORT CIRCUIT

ABOVE *Patrons who survived the fire file past fire trucks pumping water onto the three-story club. The building had no fire sprinklers or fire alarms, both of which would have saved many lives.*

On the evening of May 18, the crowd at the 3–story Beverly Hills Supper club in Southgate, Kentucky, billed as "the showplace of the Midwest," was in a festive mood. They had gathered to see John Davidson, the singer, who was to perform in the Cabaret Room.

At about 9 p.m. a busboy grabbed the microphone and told everyone to leave through the nearest exit because there was a fire. "The worst thing of all was

Fire Behavior

Structural collapse is always a danger during fires. Steel softens during an intense fire, and at 1,000 degrees a steel beam will lengthen 1 inch for each 100 feet. This will cause a roof beam fixed at both ends to buckle and collapse.

Heroes

A survivor credited comedians Jim Teeter and Jim McDonald with "saving lots of lives, including my own," by staying onstage in the Cabaret Room when the audience began to panic.

that a lot of people didn't believe me," he later said. Chief Richard Riesenberg said that his men received the first alarm at 9:15 and reached the club about 9:20.

A sprinkler system was not installed in the building when it was renovated in 1970 because Kentucky fire laws at the time did not require one, and a subsequently enacted law was not retroactive. Had there been smoke detectors, sprinklers and fire alarms, the fire might have been caught quickly.

The fabric-covered walls, carpeting, plush seating and ceiling tiles burned rapidly, pouring heavy, black,

FIRE PREVENTION TECHNOLOGY

Modern sprinkler systems are designed to automatically shower water on to a fire through heat-sensitive sprinkler heads that are placed at intervals on a piping system, usually in the ceiling area. There are four types:

The Wet pipe system has water under pressure all the time and is generally used in heated commercial spaces; the Dry pipe system has air in the pipes until fire is detected, and is used in unheated buildings so water doesn't freeze in the pipes; Deluge systems, found in places where chemical fires are a danger, such as aircraft hangers and petroleum handling facilities, release high volumes of water or foam; the Preaction system is similar to dry pipe sprinklers, but only isolated sprinkler heads go off, instead of dousing the whole area. This is used where valuables are kept, to keep the water damage to a minimum.

BELOW *Bodies of some who perished are placed on the lawn awaiting transportation to the morgue.*

The Investigation

Investigations afterwards showed that a short circuit in an electrical cord with worn insulation, on a fountain pump in the Zebra Room, caused the fire. Investigators estimated that the fire began as early as 7 p.m. in the wall, traveled up a beam and was trapped in the club's attic by the steel roof. The temperature in the room began to rise and, in fact, a wedding party was moved out of the room at 8:20 because of heat generated by the fire in the plaster wall. The occupants thought the air conditioners had failed. Thus, the warning signal of the heat was ignored.

The Reaction

The fire brought a wave of precautionary moves throughout the nation. Fire officials stepped up inspections of nightclubs for compliance with fire codes, and concerned citizens reported violations in malls, stores and office buildings. Many buildings were retroactively fitted with sprinkler systems and smoke detectors enjoyed a surge in sales.

LEFT *A technician sits in the temporary morgue in Ft. Thomas checking the possessions of victims for clues to their identification. This method of identification has remained consistent through at least three centuries. Badly burned victims have been identified by rings, watches and tooth fillings.*

blinding smoke throughout the club. The scene turned quickly into an inferno, with flames reported to be as high as 100 feet. Intense heat twisted the steel beams, and the roof caved in.

Bodies that had been removed littered the hillside surrounding the club before they were taken to the morgue. Chief Riesenberg of the Southgate Volunteer Fire Department said that most of the victims had died of smoke inhalation.

By dawn the fire had been extinguished through the efforts of hundreds of firefighters pouring tons of water on the ruins. While a 50–ton crane lifted up portions of the roof, volunteers from nearby Fort Thomas sifted through the rubble for bodies.

The town of Trenton, Ohio was stunned by the deaths of 13 elementary school employees attending a retirement party at the club. One student in Cincinnati lost his father, mother, three sisters and brother. John Davidson's musical director, Douglas Herro was killed, along with five members of the house band.

Smoke Inhalation

Smoke is an irritant to the respiratory system and even small quantities may be fatal when inhaled. Hot gases may also sear the lungs, causing internal burns. This is an extremely painful way to die, but usually one dies within minutes.

One man reported in The *New York Times,* "Everybody was pushing everybody, we did not know what to do." One survivor said he and ten others "barricaded ourselves in a room and stuffed bedding in cracks, then doused them with water."

MGM Grand Hotel & Casino, Las Vegas

DATE · NOVEMBER 21, 1980

DAMAGE · 84 DEAD, 700 INJURED

CAUSE · ELECTRICAL SHORT CIRCUIT

The death toll at the MGM Grand Hotel fire made it the second worst hotel fire in the country's history.

Most of the guests were asleep at 7 a.m. when fire

ABOVE *Smoke rises from the MGM Grand after an electrical short started a fire in the ceiling of the delicatessen. Guests were trapped on the upper floors, where there were no sprinklers or fire alarms.*

began in a delicatessen's ceiling on the hotel's main floor. An electrical short sparked the fire, which investigators later said had smoldered for about two or three

The Hotel Before the Fire

The MGM Grand Hotel was billed as the biggest, most expensive and luxurious hotel on the Las Vegas strip when it opened in December 1973. It had 2,076 rooms on 26 stories, and featured two enormous show rooms, the Ziegfeld Room, seating 800, and the Celebrity Room, seating 1,200. It had eight restaurants and the world's largest casino, holding 1,000 slot machines, 45 blackjack tables, 16 poker tables, 10 craps tables and a Keno room for 200.

LEFT *Rows of melted slot machines in the casino, after a fireball roared through it. A flashover occurs when all the contents of a room are super-heated and ignite, usually in seconds.*

FIRE-FIGHTING EQUIPMENT

Truck-mounted ladders are raised and extended to save people from tall buildings. There are generally three types: Aerial ladders, which are constructed of steel and aluminum in a truss-style design, with each section sliding out on a pulley system to make the ladder longer. They are capable of reaching heights of about 100–110 feet, but are limited in height by the angle of the ladder when fully extended and the pressure bearing down on the extended ladder when a firefighter is at the top. Tower ladders have an extending telescopic boom with a bucket mounted on the end. They can reach similar heights as the aerial ladder, but offer victims the safety of a basket rather than the rungs of a ladder. They can safely hold about 1,000 pounds. The third type is an Articulating Boom ladder, which also has a bucket, but the raising arm can be bent in the middle to lower the basket behind a wall or obstacle.

hours, but went unnoticed until it burst through the ceiling in a ball of flame. Witnesses said that a fireball rolled across the casino, melting slot machines and bursting through the front doors.

No fire alarm went off, which led to fire officials theorizing later that equipment in the basement had been destroyed. The hotel had a sprinkler system in the basement, part of the first floor and on the top floor only—the areas not under 24–hour supervision by employees.

There were no sprinklers in the casino or other guest floors. The 1970 Clark County building code did not require individual room smoke alarms. It merely required one enclosed fireproof stairway the entire height of the building.

Many guests reported being awakened by knocks on the door or the sound of breaking glass. The Fire Department was called at 7:15 and immediately began spraying water on the fire, but rescue ladders only reached the ninth floor.

Panicked guests on the upper floors fled to the roof, where more than 1,000 were rescued by Las Vegas Police and Nellis Air Force Base helicopters. Other guests were injured in the stampede for the fire stairs and in hallways.

Screaming people, trapped on the upper floors broke windows and leaned from balconies while policemen shouted, "Don't jump." At least two people did not heed their pleas and leaped to their deaths. Some guests on lower floors were able to escape by tying sheets together and climbing to safety.

Fire Behavior

Carbon monoxide is one of the most lethal gases found in a fire. The colorless and odorless gas combines with the hemoglobin in the blood to prevent absorption of oxygen. In its most severe concentration, it can cause death in one to three minutes.

By 7:30 more than 200 firefighters and police were on the scene, and they began evacuating people along the south stairway.

By 9 a.m. firefighters had contained the flames with water and fire suppressants. Then they began a room-by-room search for victims. Most of the deaths had been caused by smoke inhalation and were found between the 19th and 25th floors. Later, toxicology tests showed that all but eight of the victims died of carbon monoxide poisoning.

Paramedics and doctors worked feverishly over victims, who were spread out in the middle of the road. Survivors clad only in pajamas wandered dazed and confused along Flamingo Road until taken to the Las Vegas Convention Center, where they were given food and a cot. Fire officials set up a temporary morgue at a fire station across the road from the hotel.

BELOW *Rescuers carry out an elderly survivor. Many guests fled to the roof and were rescued by police and Air Force helicopters.*

"Many people told me that through this experience they found a new faith in God."

—Reverend Billy Graham, after a visit with victims

The Investigation

Fire chief Roy Parrish found evidence of serious code violations at the hotel, including holes that were cut into the fire walls, air circulation equipment that had been bolted in such a way that dampers could not function properly, a stairway that was shielded with plywood instead of a fire-resistant drywall, and stairwell vents that had been nailed shut.

Lawsuits against the hotel were filed for more than $2 billion, although most were later settled for about $105 million.

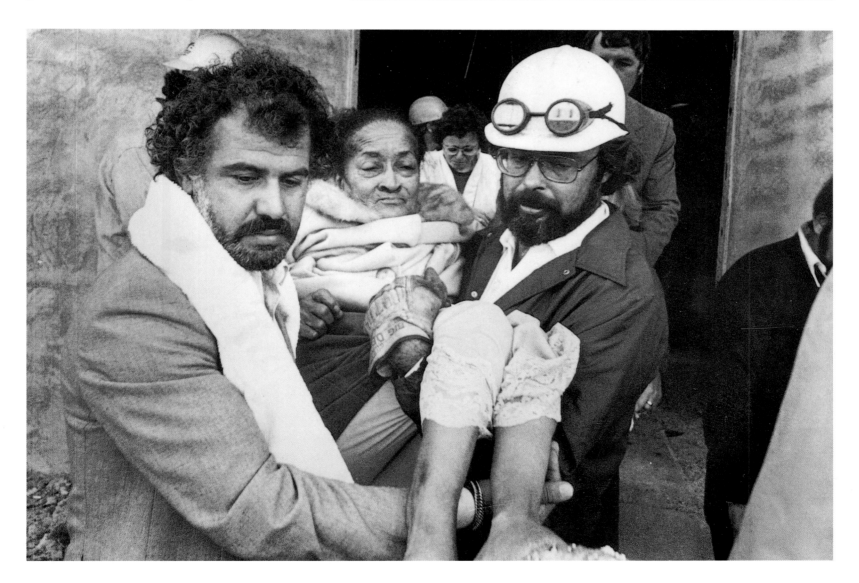

MOVE Headquarters, Philadelphia, Pennsylvania

DATE • MAY 13, 1985
DAMAGE • 11 DEAD, 61 HOUSES
CAUSE • TOVEX TR-2 EXPLOSIVE

On May 13, 500 police officers surrounded the Osage Avenue rowhouse before dawn and demanded the surrender of MOVE members. Neighbors had been evacuated the previous night.

Police tried to dislodge members by attacking their six by eight foot rooftop bunker with water cannons. Several hundred-thousand gallons of water were used to no avail. Then a 90–minute gun battle erupted, in which police fired 7,000 rounds.

Late in the afternoon a blue and white Pennsylvania

One neighboring resident was quoted in *Newsweek*, "The entire skyline was glowing red and black…it was something out of Dante's Inferno." Channel 10 cameraman Pete Kane was quoted in *Time*, "Debris was flying everywhere. Entire trees were exploding in fire." As night fell, flames shot 20–30 feet above the roofs of the attached rowhouses.

"Did I make a mistake? The answer is yes. I don't know what to say other than I'm sorry."

—Mayor Wilson Goode

State Police helicopter appeared and hovered over the roof. Two 1–pound tubes filled with Tovex TR-2 gel explosives were lit and dropped onto the rooftop bunker. Within a half minute, a powerful explosion

BELOW *Over the rooftops, flames rage through the West Philadelphia block containing the rowhouse of the radical group. Police dropped Tovex TR-2 explosives from helicopters, igniting a fire that burned at temperatures up to 7,000°.*

ABOVE *A Philadelphia police officer on a rooftop near the burning MOVE rowhouse. Firemen could not fight the spreading fire due to continuing gun battles.*

sent wood and metal flying through the air. A thick column of black smoke and flame rose into the air.

The fire quickly erupted into an inferno. Firefighters were held back because of continuing gunfire. Fire Commissioner William Richmond declared, "They are firefighters, not infantrymen." By the time firefighters were allowed to respond, 40 minutes later, the entire rowhouse was engulfed in flames and the fire had spread to three adjoining homes. The roof collapsed into the building, trapping most of the MOVE members in the cellar, where they had taken refuge.

The fire destroyed 53 houses and severely damaged eight others on two city blocks. It left 240 people homeless. By midnight firefighters using water and fire retardants brought the fire under control, although

The Cult

The radical group MOVE was an anti-technology, back-to-nature, mostly black cult founded in the early 1970s. Its members all took the surname "Africa" and their leader, John Africa, espoused a vague philosophy of anarchy, vegetarianism, and affronting organized authority. In 1978 nine MOVE members were arrested, convicted of third-degree murder and sentenced to at least 30 years in prison following a confrontation with police in which one officer was killed and several others injured.

Remaining members of the group moved into the home on Osage Avenue owned by John Africa's sister. By the spring of 1985 residents on the block were complaining to police about MOVE members stealing, verbally harassing neighbors, throwing garbage in their backyard which attracted rats, and appearing on the rooftop holding guns. They shouted obscenities from the roof with bullhorns late into the night and even jogged on the adjoining roofs.

The Explosives

For weeks and possibly months, the police had been secretly testing explosives, including Du Pont's Tovex TR-2. According to an investigation by the *Philadelphia Inquirer*, Du Pont's technical information stated that a detonation of Tovex would produce heat from 3,000–7,000 degrees and recommended that it be used underground for mining and quarrying, not in the open.

some fires continued until dawn. In MOVE's headquarters, eleven bodies were found, including that of John Africa and four children. Ramona Africa and 13–year old Birdie Africa were the only two members who escaped, by running through the flames.

A stunned nation thought that the Philadelphia police had overstepped their authority. There was an initial outpouring of support for Mayor Goode, the first African-American mayor. His promise to rebuild the burnt homes of victims by Christmas was not fulfilled and costs ballooned to ten times market value, due to poor construction and other expenses.

LEFT *Police, firemen and workers sort through the rubble on Osage Avenue. The MOVE house is at the right, where the group of men is standing. Fighting fires directly on such a large scale is nearly impossible, and at most, firefighters hope to contain the edges of the fire.*

The Investigation

An investigation in November 1985 exonerated the police for their actions. Conflicting testimony revealed the lack of coordination between city officials, the police and fire officials. In order to prevent similar situations, many cities have set up special police and fire units, recruited more minority personnel and enhanced their community-relations programs. In 1996 a civil jury awarded Ramona Africa, who had served seven years in prison, and two relatives of MOVE victims $1.5 million dollars, on the grounds the city used excessive force. She continues the fight to release imprisoned members of MOVE.

Dupont Plaza Hotel, Puerto Rico

DATE · NEW YEAR'S EVE,
DECEMBER 31, 1986–
JANUARY 1, 1987

DAMAGE · 96 DEAD,
100 INJURED

CAUSE · ARSON

More than 1,000 holiday guests were enjoying themselves gambling in the casino, laying by the pool or preparing for the New Year's Eve festivities when fire broke out in the luxurious beachfront hotel.

The fire broke out in a hotel ballroom at about 3:30 p.m., minutes after union workers rejected a contract settlement offered by management. Witnesses reported hearing a series of explosions and seeing thick smoke coming out of the ballroom. Attempts were made to enter, but a number of the doors were barred from the inside.

Witnesses said that a fireball shot across the casino ceiling, killing gamblers and croupiers where they sat or stood. The dead sat upright in chairs, as if frozen. All but a few were charred beyond recognition.

No fire alarms were sounded and there was no sprinkler system. Governor Hernàndez Colòn said that the Puerto Rican fire code did not require fire sprinklers. Smoke-filled stairways forced guests back into their rooms, where they huddled on balconies, screaming for help. Firefighters pleaded with people not to

"This was really a classic arson investigation."

—Phillip C. McGuire, Deputy Director of the U.S. Bureau of Alcohol, Tobacco and Firearms.

LEFT *Police mug shot of Hector Escudero Aponte, a disgruntled union employee, who set the arson fire by placing a lit Sterno container near plastic-wrapped furniture. He was charged with 96 counts of murder.*

FIRE FIGHTING TECHNOLOGY

Self-contained breathing apparatus (SCBA) is one of the most important pieces of equipment used by firefighters. It allows them to enter hazardous atmospheres to perform interior searches, rescues and removal of victims. It consists of an oxygen tank with mask and respirator. There are two types: Open-circuit, the most common, in which exhaled air is vented to the outside, and Closed-circuit, in which the exhaled air stays in the system for filtering, cleaning and reuse.

John Condon, who was in the mezzanine-level casino, told *The New York Times*, "There was a big burst of smoke and we went running toward one of the exits. When someone opened the door, we saw that the whole hallway was covered with black smoke. We slammed the door, went running toward the other exit and that was filled with black smoke. Then the panic began." He broke a window and jumped to the poolside patio, along with others. Sunbathers, dazed and cut by flying glass, pushed their way to the beach.

jump. Panicked guests on the upper floors, trapped by the flames, fled to the roof where helicopters rescued them and carried them to the beach.

The fire raged for five hours, until brought under control later that evening with water and fire retardants. Rescue teams wearing breathing equipment searched for survivors throughout the night, and led many to safety.

Forensic investigators from the FBI were flown to help in the identification of the bodies, and arson specialists from the Bureau of Alcohol, Tobacco and Firearms were also sent to assist.

ABOVE *The Dupont Plaza Hotel, where an arson fire was started in the ballroom. There were no fire alarms or sprinklers, and firefighters struggled for five hours to contain the blaze.*

The Investigation

On January 13 police arrested Hector Escudero Aponte, a hotel maintenance worker, and charged him with setting the deadly fire, as well as murder, arson and destruction of property. A second suspect was also arrested as an accomplice. The motive was his anger at the hotel for not agreeing to the union contract. He placed a "Sterno-type" cooking fuel can near a stack of plastic-wrapped furniture and lit it. Aponte told investigators that he did not intend to kill people, but only wanted to intimidate the hotel management. He faced a 99-year sentence for each of the 96 people killed.

King's Cross Subway Station, London

DATE · NOVEMBER 18, 1987
DAMAGE · 31 DEAD
CAUSE · OVERHEATED MECHANISM IN ESCALATOR

The worst fire in the history of the Tube, London's underground subway system, broke out at King's Cross Station at about 7:40 p.m.

The station is one of the busiest in the London Underground system, where five lines intersect, including the Piccadilly Line. Witnesses described a scene of panic and confusion. The blaze started about halfway

up a wooden escalator. Investigations after showed that the cause of the blaze was probably overheated ball-bearings beneath the escalator igniting the oil in the machine room underneath it.

The first firefighters arrived about five minutes after the fire started, under the leadership of Station Officer

BELOW *A fireman wearing a self-contained breathing apparatus descends into smoke billowing up from the Underground station. Developed in the 1970s, protective clothing, including oxygen tanks and respirators, increased the firefighters' ability to enter hazardous interiors to find and save victims and fight fires.*

The Heroes

In 1989 the widow of Colin Townsley was awarded £250,000, and he was posthumously awarded the George Medal, the highest civilian bravery award. He was among 20 London firefighters to receive commendations for their heroic actions that night.

ABOVE *Two policeman guard the escalator where the fire began. A flashover occurred, sending a ball of fire up the escalator and into the ticketing area.*

Colin Townsley, who died trying to assist a badly-burnt woman out of the station. Townsley was not wearing a breathing apparatus when a fireball came up the escalator, which acted like a diagonal chimney, drawing the flames and heat up. A "cocktail of flammable gases" resulted in a flash-over, a searing wall of fast moving flame. Some of the survivors described seeing "a sheet of flame," while others described "seeing bodies slumped against the wall, where the heat was so intense that it popped tiles and cracked solid concrete."

Trains continued to pull into the station and discharge passengers, and confused workers directed passengers onto an escalator that headed directly into the blaze.

The inquiry revealed that the King's Cross control room, which should have been the center of communications, was unmanned at the time of the fire, and much of the equipment was out of order. It was revealed that the public address system was never used during the fire and evacuation of people.

Choking smoke and intense heat made fire-fighting difficult. The heat generated was so intense that one firefighter could feel it "taking the skin off his head." Firefighters described to the inquiry afterward how, in fear for their lives, they were forced to squat or crawl among badly burned bodies while they aimed their hoses at the flames.

Sir Keith Bright, chairman of London Regional Transport, inspected the ruined station at 11 p.m. and declared that the fire was out.

Diana, Princess of Wales, joined about 400 mourners at a memorial service on December 5, held at St. Pancras Church a few hundred yards from the King's Cross station.

The King's Cross fire resulted in improvements in safety standards, changes in the command and control structure and regular fire inspections. Closer communication between fire, police and Underground officials has resulted in the Underground being much safer than it was in 1987.

The Investigation

The inquiry into the fire criticized the budget cuts at London Underground, which forced reductions in personnel at the station. It also targeted the constant repainting and thick layers of paint as a risk for creating fireballs. It cited a lack of communication between the Underground, the fire department and the police as a contributing factor in the confusion of passengers, which resulted directly in deaths.

A decision not to prosecute the managers of the London Underground was highly criticized, with many victims saying that the Director should have been held accountable.

Yellowstone National Park, Wyoming

DATES · JUNE 25–OCTOBER 1, 1988

DAMAGE · $120 MILLION; 1 FIREFIGHTER DEAD

CAUSE · LIGHTNING STRIKES

The summer of 1988 was one of the driest in recent years. The normal spring and summer rains did not fall and wildfires were burning throughout the West— 120,000 acres in Idaho, 500,000 acres in Wyoming,

---

BELOW *Tourists photograph the Yellowstone fire as flames shoot out the crowns of trees. A crown fire is one that burns across the tops of trees, frequently at a high speed and with flames that may reach up to 200 feet.*

"Even with every weapon at his disposal, man cannot always put out a wildfire."

—George T. Frampton, Jr., President, The Wilderness Society

The Free Burn Policy

The National Park Service has had a "free burn" policy since 1972. This means that, when a fire starts from a lightning strike, it is allowed to burn naturally. Environmentalists and other advocates argue that these fires clear out dry brush and dead timber. It adds nutrients to the soil in the form of ash and allows more sunlight into the forests. In fact, the lodgepole pine that is abundant in Yellowstone needs fire to reproduce because high temperatures cause their cones to burst, releasing winged seeds.

The Fan-Fire, the first of a series of lightning-caused fires, began on June 25 and burned naturally until mid-July, when it had burned over 16,000 acres. By then there was a fire on the southern boundary of the park, called the Snake Complex fire, which burned into Yellowstone Lake. Multiple large fires, including the Mink Creek Fire, which burned over 144,000 acres, the Mist Fire, which began on July 9, due to lightning, and the Clover and Raven fires, which also began in July, linked together to burn over 396,000 acres.

The North Fork Fire, the largest of the series, began on July 22, when woodcutters accidentally ignited it with a carelessly tossed cigarette. The lack of rain, combined with an 80 m.p.h. wind, whipped the flames out of control. By now, the fire was so big and moving so fast that the fire had become dangerous and the free burn policy moot. Park officials launched a massive $115 million effort to contain the fire. But by then it was too late. During periods of extreme fire behavior, which was seen frequently, direct attack and control are impossible. Over 10,000 firefighters, some from as far as Maine, came to battle the flames, but the hot, dry weather, the intensity and speed of the flames made control efforts futile. There was little the firefighters could do.

By mid-August the fires were burning 160,000 acres. The town of West Yellowstone, the areas of North Fork and Madison Junction, and the tourist complex at Canyon Village, were all threatened. They were saved through the efforts of military and civilian firefighters, who soaked them with water laced with fire-retardants. Tourism fell 30% in August, although many viewed the fire as a spectacular bonus.

By September 6th a huge wall of flame 200 feet high was approaching the Old Faithful area. Efforts were made to save the historic Old Faithful Inn. Built in 1904, it is the world's largest log building. Its majestic lobby soars 92 feet, making a perfect furnace funnel. Officials evacuated the inn and hundreds of employees and tourists fled.

388,000 acres in Montana and 2 million acres in Alaska.

Established as a National Park in 1872 by President Ulysses S. Grant, Yellowstone is one of the most famous parks in the world. In 1987 it had been visited by 2.6 million visitors. They came for the dramatic scenery and the geyser "Old Faithful", which erupts every 67 minutes.

On September 7th firefighters hosed down the inn with tons of water and fire-resistant foam. Fanned by 60 m.p.h. hurricane-strength winds, the flames swept over the Inn and surrounding buildings. When the smoke cleared, the Old Faithful Inn had survived, but 19 cabins and smaller, wooden buildings were destroyed.

Three days later, on September 10, rain began to fall on Yellowstone. The next morning, swirling snow was falling on the charred timber. Nature was extinguishing the fire, which man could not control. Small fingers of fire continued to burn for a couple more weeks, but the great fire at Yellowstone was finished.

According to National Park officials, 995,000 acres, or about 40% of the park, were burned. $120 million was spent to control the fires, but probably did not reduce the acreage burned. Because the fires burned for so long, and the price of fire-fighting was so high, the press, public and politicians questioned the National Park Service policy of letting wildfires burn. For many it is a question of how parks should be managed—should it be man or nature that decides the fate of our forests?

BELOW *A Wapiti elk stands against a smoky background at Yellowstone National Park. Animals are greatly affected by wildfires, as their natural habitat burns, driving them to other areas and removing sources of food and shelter.*

Fire Behavior

Many factors influence the spread of wildfire flames. Canyons and ridges can change fire behavior by creating wind eddies and strong upslope air movements, which can increase the speed of a fire. A fire will double in rate on a 30% slope and double again on a 55% slope. When one large fire approaches another large fire, which happened repeatedly at Yellowstone, extreme fire behavior conditions can occur. This can include area ignition, in which they burn together with tremendous speed, sometimes consuming hundreds of acres in a few minutes. Crown fires burn across the tops of trees, jumping from one tree to the next, and the flames can be up to 200 feet high.

TOP *A firefighter uses a chainsaw to cut down a partially-charred tree, taking potential fuel away from the fire to create a fireline.*

ABOVE *Ground crews who are first on the fire work alongside heavy equipment crews to construct a fortified fireline.*

LEFT *As the North Fork fire approaches the structures in the Old Faithful complex, firemen aim water onto the buildings. The Old Faithful Inn survived, but 19 other smaller buildings were destroyed.*

The Recovery

The post-fire ecosystem in Yellowstone Park has rebounded rapidly and is flourishing. Some lodgepole pines are growing at a rate of 200,000 per acre and are already over 6 feet tall. Wayside exhibits help visitors understand the complexity of a forest fire and the recovery program.

Happy Land Disco, Bronx, New York

DATE · MARCH 25, 1990
DAMAGE · 87 DEAD
CAUSE · ARSON

The worst fire in New York City since the Triangle Shirtwaist fire of 1911 happened at an illegal social club in the Bronx, on the 79th anniversary of the Triangle fire.

A Cuban immigrant, Julio Gonzalez, had arrived in the U.S. with the Mariel boatlift of 1980. He was a warehouse laborer who had lost his job six weeks earlier, as well as

ABOVE *Rescue workers crowd around the door of the illegal social club in the Bronx. There was just one front-end stairway leading up to the club, and all the windows were barred.*

his girlfriend of eight years. Lydia Feliciano, 45, worked at Happy Land. Gonzalez went to the club that night to ask her to come back to him. She refused and they argued until the bouncer threw him out. He then picked up a

The Disco Before the Fire

The Happy Land Disco, located on Southern Boulevard in the East Tremont neighborhood, had no state liquor license and had been ordered closed for fire hazards and building code violations 16 months before. There was just one front door, leading to a single stairway up to the second floor club. All the windows had been barred and boarded over. For many Hondurans in the neighborhood, the club was an after-hours disco where they could dance, drink a Honduran beer and socialize with others from their homeland.

plastic jug and said, "I'll be back." Gonzalez went to a nearby Amoco gas station and filled the jug with one dollar's worth of gasoline. He threw the gas on the floor inside the front door of the club, tossed in a couple of matches and left. It was around 3:30 a.m.

The fire became an instant torch, blocking the only means of escape. Thick, hot smoke filled the club so rapidly that victims suffocated almost instantly.

An alarm was sounded about ten minutes after Gonzalez started the blaze, and firefighters reached the

"The most horrendous circumstances I've ever seen."
—New York City Mayor David Dinkins

scene within three minutes. Flames were extinguished quickly, but all the victims were already dead. Victims fell where they stood, or still sat clutching their drinks.

The dance floor, which had been filled with people dancing to the salsa beat, was transformed into a grotesque pile of twisted bodies. Sixty-eight of the dead were found upstairs in the disco and 19 were found on

BELOW *Bodies are covered along the sidewalk in front of the club following the fire. A gasoline bomb ignited the stairs and thick hot smoke suffocated the victims.*

Smoke Inhalation

Victims usually die rapidly (within a few seconds) after inhaling smoke and gasses. Burns to the throat can lead to quick death because the airway goes into spasm, swells up and closes the air passages.

ABOVE *Julio Gonzalez, who set the fire that resulted in the worst mass murder case against one person in the U.S. He was found guilty of 174 counts of murder, as well as counts of arson and assault.*

the stairs or ground floor. Emergency crews had so much trouble removing the bodies from the upstairs that they cut a hole in the wall from neighboring construction company offices to take out the bodies.

First Deputy Mayor Norman Steisel, who rushed to the scene with Mayor David Dinkins, told *The New York Times*, "It was shocking. None of the bodies I saw showed signs of burns. They looked like wax." Some patrons struggled for the stairs and the door, but were killed trying to escape. In a twist of fate, Feliciano, the girlfriend, was one of only six people who survived the fire.

Someone erected a wooden cross in front of the charred shell of the Happy Land Club and John Cardinal O'Connor led a prayer service in English and Spanish attended by hundreds. He said, "While it is very important that we pray for those who died, we must pray in a special way for those who have survived." A memorial service for 17 of the victims was held at St. Joseph's Church in the Bronx.

On July 6, 1995, five years after the fire, lawyers for the victims' families agreed to a $15.8 million settlement. As part of the settlement, building owner Alexander DiLorenzo's three insurance companies, and 11 producers and distributors of building materials, agreed to pay $7.17 million. The insurance broker for Mr. DiLorenzo paid $200,000. And the City of New York, which faced possible liability for allowing the club to operate despite violations, paid $250,000 and agreed to waive about $300,000 in fees on settlement disbursements. The single largest payment was $6.8 million from the Scottsdale Insurance Company. It all amount-

ed to $163,000 for each life lost. It seemed small compared to other wrongful death awards, but the case had been complicated, due to questions of insurance liability. The bankruptcy of Mr. DiLorenzo, combined with the complicated subletting of the building by Jay Weiss, a real estate developer married to the actress Kathleen Turner, and the death of the club's operator, Elias Colon, in the fire, were factors in the low settlement.

The Reaction

A city-wide crackdown on illegal social clubs resulted in the closing of 14 clubs, and inspectors fanned out through the city issuing summonses for violations. From Los Angeles to Detroit, fire and building inspectors were checking underground clubs and party spaces.

The Conviction

On August 19, 1991, Julio Gonzalez was found guilty on 174 counts of murder, 2 for each victim as well as counts of arson and assault. It was the worst mass murder case against one person ever in the United States. The jury rejected an insanity plea, and after four hours of deliberation, the foreman repeated the word "guilty" 176 times. Tears of relief filled the eyes of relatives and survivors.

A Danish passenger, Leo Odeland reported, "We woke up in the middle of the night...I went out of the cabin and saw thick, black smoke. Nothing happened when I pushed the fire alarm." The staircases were not protected by fire doors, and the flames quickly enveloped the ship. Many passengers were suffocated as they tried to flee down smoke-filled corridors. Several survivors reported seeing piles of burning clothes or linen in the corridors.

Scandinavian Star Ferry, North Sea

DATE · APRIL 7, 1990
DAMAGE · 158 DEAD
CAUSE · ARSON

ABOVE *Firemen wearing Self-contained Breathing Apparatus prepare to enter the burning ship. When unknown toxins might be present, SCBA's are required equipment.*

About 500 passengers and crew were aboard the Scandinavian Star, traveling from Oslo, Norway, to Frederikshavn, Denmark, when fire was discovered at about 2:30 a.m.

A distress signal was sent out, but reports by survivors stated that by then the fire was already widespread. Crew members discovered two fires within 15 minutes that they extinguished, but a larger one was found blazing one deck below, on the car storage floor.

Hundreds of terrified passengers boarded lifeboats and were picked up by other vessels, helped by helicopters from Norway, Denmark and Sweden—340 passengers were saved.

In the North Sea, the fire was still burning more than 16 hours after it started.

The charred ship was guided by tugs to a dock in Lysekil, where the bodies were removed from the blackened 10,000–ton ferry.

The fire prompted more frequent inspections of passenger ferries and cruise ships, as well as better training of staff and crew for fire emergencies.

The Ship Before the Fire

The Scandinavian Star cruise ship had been operating under Bahamian registry out of Florida, as part of Miami-based Sea Escape Limited's fleet. It had been sold to Da-No Line of Denmark just prior to the fire. The ship had a casino, pool, disco, restaurants and shops, as well as sleeping cabins. Cars were stored in the ship's hold.

ON BOARD FIRE-PREVENTION TECHNOLOGY

Fire doors are thick metal doors, which are designed to prevent the spread of fire. They should be kept closed, but are sometimes propped open by careless workers.

BELOW *Thick smoke continued to pour out of the stricken ship after it was guided to port in Lyksekil. Fireboats shoot water from elevated hoses.*

The Investigation

An investigation concluded that the fire had been set by an arsonist, who remained unknown. It found that the ship's smoke detectors had failed, that an automatic sprinkler system did not work and that emergency generators did not work for an hour after the fire. It also found that the crew was poorly trained for fire-fighting. The Norwegian Captain, Hugo Larsen, the Danish ship owner, Henrik Johansen and the Danish ferry company director, Ole B. Hansen were all found guilty of negligence and criminal violations of ship safety laws and sentenced to prison.

Oakland/Berkeley Hills, California

DATE · OCTOBER 20-21, 1991

DAMAGE · 16 DEAD; $5 BILLION

CAUSE · EXTINGUISHED FIRE, REKINDLED

ABOVE *Firefighters battle flames that were whipped into whirlwinds by strong winds crossing the steep terrain of the canyons.*

By 1991 hundreds of million-dollar homes and apartments were nestled into the eucalyptus-and-pine-tree-covered prime real estate of the Berkeley and Oakland Hills, which were soon to become an uncontrollable inferno.

The fire that swept through the hills on that Sunday morning was the deadliest in California since the San Francisco earthquake fire in 1906. There had been a 5-year drought in Northern California and much of the vegetation was extremely dry. Many of the homeowners had not cleared their yards, gardens and roofs of dead brush, dry leaves, pine tree needles and tree limbs. This created ideal kindling, or "light fuel," when the fires reached them.

On Saturday, a small grass fire was discovered in

Oakland, which may have been caused by arson. The Oakland Fire Department responded immediately and believed they had extinguished the fire. However, during the night, the fire rekindled.

By Sunday strong winds rapidly spread the flames. The topography of hills and valleys increased the speed of the flames, forming whirlwinds and eddies where the fire crossed the canyons.

The sound of sirens shattered the quiet Sunday morning. People watched homes across the valley burst into flame and realized they too were in danger. Most residents managed to gather a few belongings and leave in their cars before their own homes caught fire.

The fire spared portions of one neighborhood, near Roble Road above the Claremont Hotel in Berkeley. The former estate, designed by Frederick Law Olmsted in 1911, contained many homes designed by such famous architects as Bernard Maybeck, Walter Ratcliff and John Galen Howard.

BELOW *One lone resident stands hopelessly with a garden hose as homes burn out of control around him. Drought conditions and dry tinder fueled the wildfire.*

One of the paradoxes of the fire was that it burned the tile and stucco Pillsbury House, which Maybeck had designed to be fireproof, while it spared his redwood Chick House. The owner of Chick House and his gardener had spent the morning wetting down the house, and the Berkeley firefighters had arrived just in time to help. Of course, it also helped that they had cleared out the garden and blown the leaves off his roof.

Others who delayed, or who believed that wetting down their wood-shake roofs would save them, were not as lucky. The firestorm blazed so quickly that some neighborhoods were surrounded by fire, leaving no escape route. At least ten people were burnt alive in their cars.

Thick black smoke blew across roads and obscured

ABOVE *A firefighter pours water on smoldering tree limbs. Thorough mopping up will ensure that fires do not rekindle once they are contained.*

LEFT *Pictured in this aerial view is one standing house, with a tile roof, that is surrounded by burned homes, in the ritzy Broadway Terrace section of the Oakland Hills. Houses roofed with tile and other non-wood materials fared better in the Oakland fire, which was the most expensive fire to date in U.S. history.*

"What showed up…as burning hot spots in that black of night, this morning were clearly the charred ruins of hundreds of homes."
—Governor Pete Wilson, after surveying the area by helicopter

vision. The fire caused the closing of two highways as well as the Bay Area Rapid Transit train. The fire came within one and a half miles of the University of California's Berkeley campus, forcing the evacuation of nearly 2,000 students.

More than 1,000 firefighters fought the blaze, while overhead 10 helicopters and 10 air tankers dropped loads of water and fire retardant chemicals. The stiff wind died down, and temperatures dropped on Sunday night. At dawn Fire Chief Phillip Lamont Ewell said the fire was contained.

Residents were able to return to the still-smoldering ruins after about three days, escorted by police or Red Cross workers.

In the hills, 2,668 houses were lost, 148 were injured and 16 died. Rebuilding began immediately, and within five years most homes had been replaced.

Persian Gulf War Oil Well Fires, Kuwait

DATE · FEBRUARY-NOVEMBER, 1991

DAMAGE · $1.5 BILLION

CAUSE · ARSON

At the end of the Gulf War, Iraq's elite Republican Guard set fire to over 500 of Kuwait's oil wells during their retreat, a vindictive move intended to deprive Kuwait of the oil that Iraq coveted. Most of the wells were blown open at the base, with master valves destroyed, allowing the oil to gush in a blaze without any control. Flames could be seen for miles, and shot more than 3 million pounds of toxic gases, soot and smoke, including cancer-causing benzene, into the air every hour.

Thick soot from the fires spread into Kuwait City, forcing cars to use headlights at midday. Residents wore pollution masks on the streets. Cows, sheep and donkeys were dying daily from the smoke, and doctors warned people that the health effects could last for years.

Three Houston-based oil well fire-fighting firms— Boots & Coots, Incorporated, Wild Well Control Incorporated and Red Adair Company—left for Kuwait in March, to help extinguish the fires.

Public Health Repercussions

Research continued on the effects of the gases and soot released by the fires. At a conference held by Harvard's School of Public Health speakers presented conflicting estimates on the number of people who would suffer life-shortening illnesses due to the smoke and soot. Some placed the number of deaths at 1,000, while others compared the pollution to that of major cities such as Tokyo and Paris, and claimed it was no worse than the pollution in those cities. Long-term effects, including possible instances of cancer, are still being monitored.

BELOW *An oil well fire explodes in a fireball near an artillery piece of the U.S. 1st Division Marines. Over 500 wells were ignited by the Iraqi forces. Oil wells have elaborate systems to stop the wells from catching fire, which the Iraqi military destroyed before setting them ablaze.*

Red Adair's company had capped 28 wells by April, but complained that the Kuwaiti government was confused and not providing guidance. A frustrated Adair, who was portrayed in *The Hellfighters* by John Wayne, was ready to pull his men, and predicted the fires could rage for five years unless the Kuwaiti's got organized.

Over $100 million worth of oil was burning every day. Soot from the fires spread around the world on the jet stream and was detected in May in Hawaii. The firefighters successfully used a combination of water and liquid nitrogen to stop the burning.

During the week of November 11 the Emir of Kuwait ceremoniously turned the last valve, which capped the last fire. It had taken only nine months to control the dangerous, toxic fires—much faster than many had expected. Kuwait had lost 94.5 billion barrels of oil, or 2% of its oil reserves. However, it expected to reach its prewar production rate of 2 million barrels a day within two years.

One year afterwards scientists found that there was less damage than expected, although it may take generations for the fragile desert ecosystem, including plants and animals, to recover from the spilled oil.

LEFT *An American firefighter at Al-Ahmadi oil field works on a gushing oil well head, still smoldering after it was extinguished. It took just nine months to cap over 500 fires.*

OPPOSITE *Oil-firefighters move toward a burning well in an attempt to cap it. Specialized oil well fire-fighting firms used water and liquid nitrogen to extinguish the flames.*

The War

The Persian Gulf War was triggered by Iraq's invasion of Kuwait on August 2, 1990. Saddam Hussein, President of Iraq, sent his troops over the border and, on August 8, Iraq formally annexed Kuwait as a province. The United Nations imposed a worldwide ban on trade with Iraq and hoped this pressure would force Iraq to withdraw. It did not, and on November 29, the U.N. Security Council authorized the use of force, unless Iraq withdrew by January 15, 1991.

The Gulf War began on January 16, with an aerial attack on Iraq, called Operation Desert Storm, led by U.S. forces. The strategic bombing runs hit Iraq's communications facilities, weapons factories, oil refineries and transportation network. By mid-February the UNcoalition, which included the U.S., its European allies, Saudi Arabia, Egypt and other Arab nations, had moved to a ground offensive, known as Operation Desert Sabre. They retook Kuwait City and moved north into southern Iraq. Iraqi forces were expelled from Kuwait, and U.S. President George Bush declared a cease-fire on February 28, 1991.

Oil Well Fires

Oil is a fuel that can ignite spontaneously in the presence of high levels of oxygen, and is very difficult to extinguish The usual technique for putting out an oil well fire is to drill a second well into the burning one and then pipe in large quantities of mud or foam until the fire is smothered. Finally, the extinguished well is sealed with a hard cap, such as metal or concrete.

Windsor Castle, England

DATE · NOVEMBER 20, 1992

DAMAGE · $60,000,000

CAUSE · HEAT FROM A SPOTLIGHT IGNITING A CURTAIN

Queen Elizabeth II called 1992 her "annus horribilis" (horrible year). It was filled with separations, divorce and scandal, and perhaps one of the lowest points was the fire which swept through the northeast corner of Windsor Castle, completely gutting St. George's Hall and the State Apartments on the 45th anniversary of her marriage.

The fire began at midday on Friday, November 20th, when the heat from a spotlight ignited curtains in the

The Castle Before the Fire

Windsor Castle was begun by William I, the Conqueror, in about the year 1070. It sits on the north bank of the Thames River on 13 acres of ground. It has two sections, or wards, with the Round Tower in the center, built by Henry II. The Lower Ward, or West Ward, contains the Albert Memorial Chapel and St. George's Chapel, the chapel of the Order of the Garter. It was built in the Perpendicular Gothic-style by Edward IV and was completed in 1528. The Upper, or East Ward, contains the private apartments of the monarch and State Apartments for visiting dignitaries. St. George's also contains the Waterloo Chamber and St. George's Hall, built in the 14th Century. St. George's Hall is an ornate 185 foot-long room used for state dinners.

Private Chapel. The flames roared through the State Apartments and moved swiftly into St. George's Hall, which was lined with the flags of the Knights of the Garter. The ornately carved wooden roof completely collapsed within the charred walls. Workers frantically moved furniture, paintings, carpets and tapestries into the courtyard, to save the valuable pieces from the flames.

During the night, flames could be seen shooting up from the castle, until firefighters contained it by dawn, using thousands of gallons of water. Queen Elizabeth trudged through the rubble during a rainstorm on Saturday to inspect the damage.

The castle was uninsured, and after initial questions about who should pay for repairs, 30% was paid by the government and 70% was paid by the monarchy.

LEFT *Workmen load some of the art treasures from Windsor Castle onto a truck following the fire. Because of the quick reaction of the firefighters and military only a few pieces were lost, including a large sideboard and an oversized painting.*

OPPOSITE *Flames and smoke billow from the windows and roof of the medieval castle. The heavy-timbered roof of St. George's Hall collapsed.*

The Restoration

The restoration was the responsibility of two committees: the restoration committee chaired by the Duke of Edinburgh and made up of conservationists, and the design committee chaired by Prince Charles, made up of architects. The debate between modern and classical architecture raged in the press, with some architects advocating glass and steel while more conservative architects proposed a "gothic-style" restoration. In May 1994 the firm of Sidell Gibson was chosen for the restoration, which was done in a classical style.

Five years to the day after the fire, the restoration was complete. Over 1,500 craftsmen were employed to carve wooden gargoyles, weave new silk damask wallhangings, create new curtains and recreate plaster decorations. Over 500,000 sheets of gold leaf were applied to the plaster decorations. The ceiling of St. George's Hall required 350 mature oaks to form the beams, which each weighing over a ton.

"My reaction was shock and horror at the fact that it took hold so quickly."
—Prince Andrew, His Royal Highness, the Duke of York

A new stained glass window, by Joseph Nuttgen, in the Private Chapel, honors the valiant firefighters by depicting St. George flanked by firefighters battling the flames.

The Queen and Prince Philip's 50th anniversary was celebrated in the new hall on the day of its completion, with a royal ball, including seven Kings, ten Queens, 26 Princes and 27 Princesses.

BELOW *An aerial view of the damage, showing the roof collapse. Nine rooms were damaged and more than 200 tons of debris removed.*

Malibu/Laguna Beach Brush Fires, California

DATE · OCTOBER 27–
NOVEMBER 8, 1993
DAMAGE · 3 DEAD; 200,000
ACRES; $1 BILLION;
200 INJURED,
INCLUDING 186
FIREFIGHTERS
CAUSE · ARSON

Fast-moving fires driven by fierce Santa Ana winds swept down the canyons from the Santa Monica Mountains, northwest of Los Angeles, to the Pacific Coast.

ABOVE *Flames burn up a hill toward a home in Malibu on Tuesday afternoon, November 2. The canyons acted like funnels, forcing flames through their narrow walls at speeds over 100 miles an hour.*

Six years of drought had left a large accumulation of dead trees and dried shrubs in the hills surrounding Los Angeles. From Ventura County to Laguna Beach, and south to the Mexican border, the fires started by lightning and arson ignited the brown chaparral brush.

ABOVE LEFT *A firefighter is silhouetted against a wall of flames in Malibu on November 2. Santa Ana winds barreled down mountain canyons into the seaside enclave, forcing thousands of panic-stricken residents to flee. Firemen were able to save some homes, but many others were lost.*

BELOW LEFT *A home is engulfed in flames in Malibu on November 2. Fierce Santa Ana winds swept down the canyons to the coast.*

Canyons became fire corridors, through which flames roared with temperatures high enough to melt metal. Heat tornadoes drove flames right to the ocean.

The first flame was reported on November 2 at 1:19 p.m., off the 16th tee of a golf course in Thousand Oaks, north of Malibu. Arson was suspected, and this fire eventually destroyed 35 homes on 35,000 acres. The next morning, 60 miles east, a homeless man lit a bonfire, which by noon had charred 4,000 acres and burned hundreds of houses in Altadena. Copycat arsonists ignited fires near Irvine and in Riverside County. By Wednesday, November 3, there were 100 fires raging throughout the state.

A swath of fire cut through Laguna Beach, destroying 16,684 acres, 318 homes and forcing the evacuation of the entire community. Flames shot 70 feet in the air and made 50–foot jumps across canyons and roads.

Firefighter Injuries

Over 150 firefighters were injured dealing with the steep terrain, rattlesnakes (one man was bitten), soot, smoke and treacherous shifting winds that whipped the flames back and forth along canyon walls.

"You don't stop worrying about a fire in Malibu until a good three days after they put it out."

—Rod Steiger, actor, Malibu homeowner

Million dollar homes seemed to explode, and palm trees ignited spontaneously. Flakes of ash fell in downtown Los Angeles.

Over 6,000 firefighters from around the state, 765 engine companies, and six fire-retardant-dropping air tankers fought the fires, using 100–gallon buckets refilled from the Pacific Ocean. But there was little they could do to control the blaze because of the extreme heat. In addition, a lack of water pressure from low and overused reserves forced them to use swimming pools for water.

The authorities ordered the evacuation of most of Malibu, including Malibu Colony, where many celebrities have homes. Actor Sean Penn's home was destroyed. With the wall of fire just three miles away, workers at the J. Paul Getty Museum in Malibu raced to protect the museum's multi-million dollar collection. Much of the collection was moved to the basement, the grounds of the building were soaked with water and staff members were prepared to cut down the trees surrounding the building. A tactical defense of cutting huge fire lines around nearby highways mounted by firefighters saved the museum in Malibu. Museum curator John Walsh said firefighters tackled the flames with help from helicopters and C-130 cargo planes that dumped water and fire-retardant chemicals

---

BELOW RIGHT *This special fire fighting helicopter is equipped with a bucket suspended by cable from the cargo hook. The helicopters hold substantially less than airtanker planes, however, they are much more accurate and very effective at extinguishing small spot fires that often develop around larger fires.*

to reinforce the firebreaks on the museum and grounds.

Officials declared the wildfires under control on November 4, three days after the fires had started. The damp ocean air and calm breezes had helped to sap much of the strength from the ferocious blaze. It is believed that 19 of the 25 fires which ravaged the Los Angeles area were caused by arson.

Exhausted firefighters were able to take breaks in

Firefighter Tributes

Following the deadly fire that swept through the hillsides, many people returned to the seaside paradise and found their homes intact. They responded with an emotional tribute to the 6,000 firefighters from around the West who fought this blaze house by house.

"GOD BLESS YOU!" read a hand-painted sign along Pacific Coast Highway. Dozens of similar signs adorned houses. People blew kisses at fire engines rumbling along streets. Groups fed the thousands of firefighters at a base camp, shaking their hands and singing their praises. Thomas Laskins, 47, handing out bottles of fresh orange juice to one crew, said, "They own this town." Celebrities, including actor Michael Nouri, hand-delivered flowers, cookies and bagels. Families startled grimy, stubbly-faced firefighters by walking up and introducing their children, insisting firefighters visit their homes.

"They've been incredibly kind and appreciative," said Roosevelt Henderson, 33, a firefighter from Alhambra, California. He and other members of his strike force spent the day at a home in the same canyon where they spent the previous night putting out fires.

The Saved Museum

The J. Paul Getty Museum in Malibu, opened in 1974, is a reproduction of an ancient Roman villa, the Villa dei Papiri at Herculaneum, destroyed by the eruption of Mount Vesuvius in A.D. 79. It has one of the strongest collections of Greek and Roman art in the country. Works by the Old Masters, French furniture and drawings were moved in 1997 to the new Getty Center, atop a hill in Brentwood, designed by Richard Meier.

Damages to selected towns:

Altadena: 5,700 acres, 118 homes, $58.5 million

Laguna Beach: 16,684 acres, 318 homes, $270 million

Malibu: 18,000 acres, 350 homes

Thousand Oaks: 39,000 acres, 43 homes

Winchester: 25,100 acres, 30 homes, 2 farms, $2.6 million

12–hour shifts. Most crews went 24 hours without sleep, and some went as long as three days. "We eat and wait, eat and wait," groaned Mark Henderson, 40, a Grand Teton National Park ranger. They also swapped fire stories, including the great Oakland/Berkeley fires of 1991. "It's a strange feeling," acknowledged Roger Bird, an Oakland fire captain. "I can feel people looking at us as though we have some special understanding of fire."

Due to the successful evacuation efforts, only three people were killed by the fire, an elderly couple in a pickup truck and British-born filmmaker Duncan Gibbons.

Los Angeles Mayor Richard Riordan apologized later for failing to show sympathy and support for victims of the Malibu wildfires, who live outside the city of Los Angeles, or appreciation of the firefighters from throughout the state, who waged war against the flames.

A week after the firefighters put out the last of 26 major wildfires that had scorched Southern California, residents and work crews prepared for the winter rains that usually cause floods and landslides. The fire-blackened hills were reseeded, trenches were dug and sandbags filled.

BELOW *A single home sits virtually untouched in Laguna Beach after wildfires reduced neighboring homes to rubble. Odd wind currents, tile roofs and a lack of trees and bushes help explain why a house is spared amid the ruins.*

Gran Teatro del Liceu, Barcelona, Spain

DATE · JANUARY 31, 1994
DAMAGE · $95,000,000
CAUSE · SPARK FROM A
WELDER'S TORCH

The Liceu fire of 1994 was ignited by a spark from a welder soldering a piece of the iron curtain, a metal stage curtain used to contain an accidental fire on stage. The flames spread rapidly because of the plaster and wood construction of the building. The plush velvet seats and curtains added fuel to the fire, as did the scenery and backstage flats. The fire quickly destroyed the entire auditorium and stage area, leaving only the outer walls, foyer and main staircase standing. Fortunately, nobody was killed. Catalans rushed to the Ramblas promenade and wept aloud at the loss of their beloved opera house.

The 8.7 million pesetas insurance on the theater would not cover the cost of rebuilding, and the hereditary private owners could not afford to do it without government help. After complicated negotiations, the seat holders agreed to give the ownership of the theater to a public agency.

After three years and $95 million, the Gran Teatro del Liceu was opened by King Juan Carlos for the 1999–2000 season with a performance of Puccini's *Turandot*, which had been the next opera scheduled to play in 1994, when the fire occurred.

"The reconstruction effort began the very day of the catastrophe with everybody coming together to make it possible."
—Josep Caminal, General Director, Gran Teatro del Liceu

The Theater Before the Fire

The building of the Liceu was begun in April, 1845, and the opera house opened on April 4, 1847. It is located on the famous Ramblas, in the heart of Barcelona. It was designed by Miguel Farriga I Roca, and followed the Italian model, with a horseshoe-shaped auditorium seating 4,000 people. The initial construction was funded by the Catalan bourgeoisie, who received a number of seats in perpetuity depending on the size of their donation. The theater burned down in 1861 and the private owners rebuilt it in just a year. It survived a bomb attack by separatists, which killed 20 people in 1893, during an opening night performance of *William Tell*. By the 1980s, the theater still featured performers such as Jose Carreras, but the supporting casts were mediocre, and the staging areas were cramped and old-fashioned. In 1987, a committee of opera lovers, businessmen and architects proposed a renewal plan for the Liceu. After many delays, construction was begun.

ABOVE *The burned-out auditorium and stage area of the opera house. The fire left only the outer walls, foyer and main staircase standing.*

LEFT *A tower of thick smoke billows from the 150–year old opera house, which burned accidentally after a spark from a workman's blowtorch ignited a stage curtain. Theaters today offer more exits for patrons, as well as fire alarms and sprinklers.*

The Rebuilding

Neighboring building owners were persuaded to sell their properties to allow expansion. Josep Caminal oversaw the restoration, adroitly balancing private and public partnerships. An architectural firm led by Ignasi de Sola-Morales performed the restoration. The horseshoe shaped auditorium, with its five rows of balconies and gilt ceiling decorations, was recreated. The stage area was expanded and now has state-of-the-art capabilities, such as audio-visual hookups.

Glenwood Springs, Colorado

DATE · JULY 6, 1994
DAMAGE · 14 FIREFIGHTERS DEAD
CAUSE · LIGHTNING STRIKE

The fire probably started with a lightning strike on a juniper or spruce tree sometime over the 4th of July weekend, about five miles west of Glenwood Springs, a town of 6,500 residents, 60 miles west of Vail. It started as a relatively small fire, burning about 50 acres on the side of Storm King Mountain. Rural development had expanded the town up the mountain, and homes and roads were right in the fire's path.

On Wednesday the 6th, a group of firefighters was assembled to combat the persistent flames. It included a group of elite Federal Fire Jumpers named the Hot Shots, from the Prineville, Oregon, area, who were helicoptered to the site.

A second group of fire jumpers and three local fire companies that walked in from the road joined them. There were a total of 52 firefighters on the blaze.

Most of the morning was spent cutting a fireline, or cleared zone, face to face with the flames. "Hand crews"

ABOVE *Firefighters preparing to move into a field.*

The Hot Shots

The Hot Shot program was created in 1960 as a fast-response network, to operate side-by-side with the Smokejumpers, a Federal corps of firefighters who parachute in behind the lines of forest fires.

Fire Behavior

Fire will usually burn faster and more intensely going up a slope because the flames are closer to the ground fuel. Heat rises along the slope causing a draft, which increases the speed of the fire. Narrow canyons can create a chimney effect.

The Investigation

Speculation and questions about the way the fire was managed came from parents of the dead as well as other firefighters. A Federal investigation into the blaze found that firefighters and supervisors made crucial errors in judgment and violated safety procedures. The report found that "escape routes and safety zones were inadequate," and that firefighters "were confused about who was making the decisions on strategy and tactics." It also noted that "several firefighters played heroic roles." In one instance, a firefighter pulled a woman from the ground, forcing her to get up and keep running. Still, the report cited a lack of communication, and found that meteorologists were not consulted. It said that the "can-do attitude of supervisors and firefighters led to a compromising of standard safety procedure, like a failure to plot an escape route."

Fire Fighter Safety

The essential elements of wildland fire-fighting safety are included in the acronym LACES: Lookouts, Awareness, Communications, Escape Routes, Safety Zones. The lead fire fighting personnel are responsible for advance-briefing all firefighters on escape and safety plans and providing reliable communications between the firefighters. Every firefighter digging a fire line or fighting a fire must know exactly where to run in case the blaze approaches too quickly or changes direction. The carefully assigned escape routes and safety zones are defined by the terrain and natural firebreaks, i.e. places where the fire definitely will not go. This can include already burnt out areas. Sometimes, due to erratic weather, the safety zones change, so communication of weather fronts and changing fire conditions is essential for protecting the lives of the firefighters.

ABOVE *Smoke billows from the fire on Storm King Mountain where 14 firefighter "hotshots" lost their lives They were caught in a ravine and tried to outrun a firestorm that was estimated to be traveling 100 feet per minute.*

usually work in rugged terrain, where bulldozers cannot be used to clear trees and underbrush. Firefighters typically use handtools, such as chainsaws, rakes, axes, hooks, shovels and hoes to clear flammable debris, such as grass, pine needles and trees from a line between the fire and the town.

A FIREFIGHTER'S LAST RESORT

A special heat shelter is a last resort for firefighters. It is a thin metal-like covering that wraps like a sleeping bag or pup tent around the firefighter while he or she is lying face down on the ground. It can reflect heat for short periods of time as a fire moves onward, but it can not withstand a slow moving fire that lingers overhead. They work well in fast moving grass fires, but are essentially worthless in superheated, crowning wood fires. It should be placed on a flat, cleared area, free of debris that could ignite. Inside the shelters, firefighters can survive heat over 180 degrees.

Some of the Hot Shots had moved into a canyon, digging fire lines. This placed them precariously above the fire, but below the top of the mountain. Suddenly, at 3:30 p.m., the wind changed direction and began gusting in excess of 50 miles per hour. With the wind fanning the flames and redirecting the fire to unburned fuel, the fire exploded into a superheated blowup that jumped the ravine, trapping firefighters between two walls of flame, near the top of the ridge. Many of the firefighters raced the fire up the side of the mountain in an excruciating run, hoping to get over the top and find shelter on the other side. Others looked for bare ground and pulled out their survival shelters. Other firefighters ran straight through the flames to the ravine bottom where the fire had already burned. When the firestorm had passed over the slope, 38 of the 52 firefighters had survived. The fire grew to 2,000 acres and lasted five hours. Late on Thursday, July 7, rescue teams recovered the bodies of four female and eight male firefighters. The remaining two men were brought out on Friday the 8th. Some of them had been found in their shelters, which had failed due to the prolonged exposure to the flames. Nine members of the Prineville Hot Shots, five of the men and all four women, had died.

BELOW RIGHT *After carefully assessing a safe landing spot, the team leader gives the signature tap on the calf to indicate "jump" and out of the plan they sail.*

BELOW *A firefighter uses a Pulaski, which is a handtool that combines an axe and pick. It is one of the most important tools for forest firefighters and is used to construct firelines.*

BOTTOM *Fire jumpers check their gear and prepare to board the plane. Teams of "hotshots," or highly trained professional firefighters, parachute into remote areas extinguish small forest fires before they grow out of control.*

"This is a great tragedy involving the finest fire fighting crews this country has ever produced."
—Bruce Babbitt, U.S. Interior Secretary

Teatro La Fenice, Venice

DATE · JANUARY 29, 1996
DAMAGE · $100,000,000
CAUSE · ARSON

The gilded opera house, which had premiered Verdi's *La Traviata* and *Rigoletto*, was destroyed in a dramatic fire that left it a smoldering ruin.

La Fenice had been closed in August 1995 for a renovation, and was empty at the time of the fire. Fires in three different areas were discovered at about 9 p.m. and firefighters responded quickly. However, the two canals on which the theater was situated had been drained of water as part of a dredging project. Venice's famed fireboats could not gain access, and firefighters were forced to piece together hoses to pump water from the Grand Canal.

The fire, with its suspiciously strong start, quickly spread, and after the roof collapsed 150–foot flames shot into the Venetian night sky. Firefighters used helicopters to dump water onto the fire and prevent it from igniting neighboring buildings. The fire lasted nine hours and gutted both the auditorium and ballroom sections of the interior.

The Rebuilding

Mayor Massimo Cacciari vowed the next morning that it would be rebuilt "dov'era, com'era [as it was, it will be]." The Italian government pledged $12.5 million toward the reconstruction, and other pledges of aid came from all over the world. The debate about whether to restore the building to its 18th Century appearance or insert a modern building raged in the press for months. La Fenice has not been reconstructed, due to judicial investigations and political squabbling.

The Theater Before the Fire

Built in 1792, following a design competition won by Giannantonio Selva, La Fenice was built in a Neoclassical style. It had five levels of private boxes that enabled rich Venetians to see and be seen. The auditorium section burned to the ground in 1836, but the front ballroom was only slightly damaged. Reconstructed by the Meduna brothers, the theater rose from the ashes in a year, like the legendary Phoenix for which it was named. There was a complete restoration by Eugenio Miozzi in 1937.

RIGHT *An aerial view of the opera house and its collapsed roof. When superheated, deteriorating exterior masonry walls, supporting heavy wooden roof beams, led to a structural failure.*

FOLLOWING PAGE *Three days after the fire, water continues to be sprayed by firemen onto the smoldering ruins. Heavy wooden timbers are often termed "long burners" because they typically will smolder and even re-ignite.*

FIRE-FIGHTING TECHNOLOGY OF THE TIME

A fireboat is a massive floating fire engine and is a very effective fire-fighting tool because it can pump huge quantities of water onto a fire. Most land-based fire engines have a water pumping capacity of only 1,000 gallons per minute, while fireboats can pump 7,000 gallons per minute. The water supply is drawn directly from the lake or river in which they are floating.

Chapel of the Holy Shroud, Turin, Italy

DATE · APRIL 11-12, 1997
DAMAGE · $60 MILLION;
　　　　　　PARTIALLY
　　　　　　DESTROYED
CAUSE · UNKNOWN

The fire in the Chapel of the Holy Shroud, where the Holy Shroud of Turin is kept, broke out shortly after the end of a dinner party to celebrate the upcoming reopening of the restored chapel, in the adjoining Royal Palace Museum. The party was attended by United Nations General Kofi Annan, former Prime Minister

BELOW *Firemen carry out the Holy Shroud of Turin. They rescued it by smashing through four layers of bulletproof glass.*

The Building Before the Fire

The Chapel of the Holy Shroud is regarded as one of the finest Baroque buildings in Europe. The domed chapel, designed by Guarino Guarini, and built from 1657–1694, rises behind the Cathedral of San Giovanni, built in 1491–1498. The Holy Shroud is located behind the main altar of the cathedral, in an elevated position, which is reached by symmetrical stairs. Its circular plan is dominated by the elaborately carved reliquary holding the shroud, designed by Antonio Bertola. In 1825–26 a large glass partition was placed between the cathedral and the chapel by King Carlo Felice to separate the common people from the chapel, where the royal court worshiped and controlled access to the shroud. Immediately prior to the fire, the Chapel had undergone a three-year restoration effort and was to reopen in three weeks.

The Restoration

The Italian government pledged $60 million dollars for restoration of the Chapel and Palace. Plans included the installation of more smoke and fire alarms, and additional guards. Successful reconstruction of the Chapel will depend on fragments of charred stucco and marble being salvaged and the skill of restorers in copying Guarini's details.

"God gave me the strength to break the glass."

—Firefighter Mario Trematore, who rescued the shroud

The Holy Shroud

The Shroud of Turin is a sepia-toned impression of a man bearing the wounds of Christ. The image is believed to have been transferred to the cloth when Nicodemus and Joseph of Arimathea carried the body of Christ, covered by the cloth, to his tomb. The shroud was given to the Dukes of Savoy in the 15th century, and was brought to Turin in 1578.

Rescuing the Holy Shroud

Firefighter Mario Trematore hammered through four layers of bullet-proof glass to rescue the Holy Shroud in its silver box, and was given a special papal award by Pope John Paul II. This was the second time the shroud had been rescued from fire—in 1532 it narrowly escaped destruction while at Chambery, before being brought to Turin.

Giulio Andreotti and Turin's mayor, Valentino Castellani, among other dignitaries. Drinks had been served in the chapel before guests moved to the palace for dinner. At 11 p.m. the fire alarms went off, but museum personnel were unable to find evidence of the fire for a half hour.

Finally, as guests were leaving, flames were seen roaring out of the upper windows of the chapel. Firefighters were then called in, but due to the narrow streets and crowds surrounding the building, it took about a half hour for them to arrive. Firefighters fought for hours to control the blaze, but the interior of the chapel was gutted. The 19th century organ of the Cathedral was destroyed and some of the archives were damaged. The west tower of the museum, formerly the home of the Dukes of Savoy, collapsed, and 85 minor works of art were lost in the blaze.

Tropical Forest Fires, Indonesia

DATE · AUGUST-DECEMBER, 1997

DAMAGE · 1,000+ DEAD; 7.4 MILLION ACRES

CAUSE · MANMADE, LAND-CLEARING FIRES

One of the worst ecological disasters in the world blanketed a large swath of Southeast Asia in a choking haze of smoke for months. Singapore, the Philippines, Brunei, Indonesia, Malaysia and southern Thailand were directly affected by the fires burning on the Indonesian islands of Sumatra and Borneo.

The fires were mostly intentionally set by Indonesian and Malaysian logging companies that clear cut the forest to burn stumps and clear the land for oil palm plantations. Vast tracts of jungle were set ablaze.

In Sumatra, over 200 fires were burning by August, and by September 650 fires were raging in the province of Kalimantan. There was no effort by the government of President Suharto to fight the fires, possibly because many of the companies were owned by relatives of government officials.

In the mad rush towards modernization in Southeast Asia over the past 30 years, environmental concerns have taken a back seat to the desire for wealth and economic growth. Since the 1960s, Indonesian forests have become a substantial source of state revenue and a means to spread political and social control over the vast archipelago.

BELOW *Flying through a smoky haze, a U.S. C-130 Hercules water bomber from the Wyoming National Guard drops 3,000 gallons of water on burning forest fires on Mount Arjuna in Batu, 400 miles east of Jakarta. Washington extended the waterbombing program to fight the choking haze in the region.*

El Niño

El Niño begins when the trade winds, which usually blow across the Pacific Ocean from East to West, following the equator from South America to Indonesia, diminish. This causes a bulge of warm ocean water, that the winds usually hold in place near Indonesia, to move back towards South America. This movement of warm water affects weather worldwide. The jet streams shift north, and thus fewer storms reach the U.S. Northern Plains, Ohio Valley and the Northeast, causing milder winters. But, throughout the southern U.S., winters are wetter in an El Niño cycle. Countries like Chile experience increased rain, and Peru gets more snow in the mountains. In addition, plankton and fish near South America starve because the cold water that contains nutrients cannot rise to the surface, due to the warm water. An El Niño cycle usually occurs once every 5–7 years.

"The sky in Southeast Asia has turned yellow and people are dying. What we are witnessing is not just an environmental disaster but a tremendous health problem being imposed on millions."
—Claude Martin, Director General, World Wide Fund for Nature, Geneva

BELOW *A woman carrying her child watches helplessly as raging wildfires engulf her neighborhood in the slums of south Jakarta. Wooden structures fed the flames.*

The fires were accelerated by a drought brought on by El Niño. An extended dry season, caused by the warm ocean currents of El Niño, compounded the problem by making the fires easier to start and harder to put out. It also delayed the arrival of the autumn monsoon rains, which would have extinguished the fires.

The smoke from the fires combined with urban pollution in the cities of Southeast Asia. In Kuala Lumpur, Malaysia, the sun was blocked out for weeks. The Pollution Standard Index hovered between 200–300, a range that is considered "very unhealthy." In the U.S., warnings are given when the index reaches 100. People hurried along the streets of Kuala Lumpur wearing surgical masks, and newspapers instructed people to drink plenty of fluids. Millions of people throughout the region were coughing and wheezing. In the Malaysian state of Sarawak, on Borneo, the index climbed to 839, far above the level that can kill. A state of emergency closed schools and businesses, and visibility was at an arm's length. Experts said that breathing the haze was as dangerous as smoking 80 cigarettes a day.

Wildlife was also affected. Tigers and elephants fled the burning jungles. Birds fell from the skies. Hundreds of terrified orangutans were slaughtered as they sought shelter from the fires in villages and palm oil plantations.

Tourism throughout the region was also affected. Beaches from Thailand to the Philippines were shrouded in haze, and hotels, restaurants and retailers in many countries were complaining of falling trade.

The fires affected agriculture by blocking sunlight,

ABOVE *A vast area of what used to be rainforest is now ashes, after a huge wildfire near the city of Palangkaraya, on the island of Borneo in Indonesia. The smoke from the fires combined with urban pollution in the cities of Southeast Asia, choking people and causing widespread devastation.*

which slowed the growth of fruits and vegetables and reduced the yields of corn and rice. Food shortages and starvation faced many inhabitants, as well as the forest animals. Bees and insects disappeared in many areas, disrupting pollination.

The use of fire to clear the land in Southeast Asia continues today. Each year more tropical forest is burned and more hardwood is lost. The ecological consequences for global warming are being studied, but the governments of the region continue to grant permits for logging and development. The fires burrowed deep into the peat bogs and coal seams, where experts predicted they could smolder for years.

Air Pollution

According to the World Health Organization, at least 2.7 million people die each year from illnesses caused by air pollution. Below are some of the countries with the worst incidences of pollution-related deaths:

REGION	DEATHS
India	673,000
Sub-Saharan Africa	522,000
China	443,000
Other Asian countries	443,000
Latin America	406,000

Los Alamos, New Mexico

DATE · MAY 4-20, 2000
DAMAGE · 405 HOUSES,
$1 BILLION
CAUSE · CONTROLLED BURN
JUMPING FIRELINES

FIRE FIGHTING TECHNIQUES

Backfiring is a special fire-fighting technique for wildfires, requiring extensive planning. Firefighters start from an anchor point, such as a stream, road or man-made firebreak a short distance from the main fire, and purposefully set another fire between the firebreak and the main fire. Draft winds created by the main fire pull the backfire into the main fire. A backfire is set to eliminate fuel for the main fire, incorporate smaller hot spots, and provide safety zones and escape routes.

The Los Alamos fire was purposely set by the U.S. Forest Service as a controlled burn on Thursday, May 4 in the Bandelier National Monument.

The conditions at Bandelier seemed right for a controlled burn, according to the superintendent, Roy Weaver. He believed the temperature was right, the humidity high enough and the wind sufficiently low. He claimed that weather reports had been received throughout the day that did not predict strong winds. However, over the weekend, fierce winds gusting to 45 miles per hour whipped the flames into a firestorm that jumped firebreaks that had been prepared to stop the fire. The fire leaped into the forest canopy and jumped from tree to tree, blackening hundreds of acres. The pine trees literally exploded when the sap was superheated.

On Sunday, May 7 public safety officials ordered the evacuation of 11,000 people from the town of Los Alamos. Police and fire officials went door-to-door requesting that residents pack and move out. The fire, called the Cerro Grande fire after the mountain where it began, moved towards the east and south, into town.

By Thursday, May 11 the fire, driven by strong winds, had destroyed at least 260 homes and was threatening the Los Alamos National Laboratory, where nuclear weapons are designed. Smoke from the fire rose 20,000 feet into the air and was blowing into Texas. Helicopters dumped huge scoops of water from the air, while more than 1,000 firefighters and emergency workers fought the fire on the ground by enlarging the firebreaks, extinguishing smaller fires and setting backfires.

Many residents said they only had time to throw some clothes in a bag, grab their children and flee. The flames burned houses and cars indiscriminately. Some homes were charred ruins, while neighboring houses were untouched. Varying wind conditions, and cleared

RIGHT *Homes in West Los Alamos are engulfed in flames as a controlled burn set by the National Park Service in Bandelier National Monument spread out of control. Whirlwinds of fire, traveling at high rates of speed, forced the evacuation of all 18,000 residents.*

The Investigation

On Thursday, May 18 U.S. Interior Secretary Bruce Babbit told a news conference that, "the calculations that went into this were seriously flawed." A Federal investigation found that almost every aspect of the plan for the prescribed burn was poorly conceived.

The major conclusions were that they did not take into account the complexities of the terrain and the material to be burned. No one analyzed the plan with a critical eye toward unexpected developments, which resulted in ignoring the need for additional firefighters and equipment if the fire went out of control. Wind predictions for the days following the ignition were not taken into account. Coordination among federal, state and local agencies was lacking and firefighting techniques did not comply with federal standards. Nobody, as of yet, has been indicted.

A Controlled Burn

A "controlled," or "prescribed," burn is a fire that is purposefully set to clear dense brush that has accumulated, in a preemptive effort to avoid a much larger natural fire. In deciding a planned fire will remain under control, several factors are used to create a computer model: the quantity of underbrush, dead wood and tinder in the targeted area; the moisture in the fuel; the temperature and humidity of the air; and the wind velocity. "It's never completely safe," according to Dr. Wallace Covington, a professor at Northern Arizona University's School of Forestry in Flagstaff, who told *The New York Times,* "You offer up a little prayer and hope for the best because the situation might get out of hand."

The Reaction

In August 2000 the Clinton administration unveiled a revised approach to fire-
fighting. It continues the policy of prescribed fires and additionally offers a plan
to thin out forests of small trees and underbrush using machinery and manpower. Congress approved $1.7 billion in new financing for federal firefighting agen-
cies. The money will allow the Forest Service, the Bureau of Land Management,
and the National Park Service to buy more fire engines, helicopters and tanker
planes, and hire more firefighters, smoke jumpers and support workers. And
local fire departments will be able to apply for federal grants to buy more
engines, hoses and radios to improve firefighting efforts in remote areas.

Over the past fifty years, the Forest Service policy of suppressing small fires
on public lands, which had begun in 1910 after the "Big Burn" fire in Montana
and Idaho burned about three million acres, suppressed the natural cycle of
fires that cleared the excess vegetation, leaving forests prone to future fires.

The hope is that in time these new techniques of thinning and prescribed burns will make the nation's forests less dense and less prone to disastrous large wildfires.

ABOVE *Well-developed columns of smoke rise
hundreds of feet into the air, indicating intense
burning. Firemen could do little, due to the intensity and speed of the fire.*

ground surrounding homes, sometimes made the dif-
ference in whether the home burned or not. The strong
winds pushed the blaze onto the grounds of the
National Laboratory. Some 40 trailers and outbuildings
were burned, but the lab's main buildings were saved by
firefighters' efforts clearing the area of light fuels and
watering the building and surroundings. Energy
Secretary Bill Richardson said that the plutonium and
other radioactive materials were contained in concrete
bunkers built to withstand a direct fire hit, but he felt
that extra precautions were necessary to ensure environmental safety.

The fire swept out of Los Alamos and continued
onto the towns of White Rock, Abiqui and the Espanola
Valley. An additional 15,000 people had to evacuate
their homes and 400 additional homes were lost.

"Right now we should be focusing on doing everything we can to minimize the damage of the fire and protect the lab assets, deal with the human problems."

—President Bill Clinton

BELOW *A helicopter has transported firefighters to a remote hillside. The "helitack" crew is trained to work in remote locations using hand-tools to fight spot fires and establish fire lines in areas that heavy equipment cannot reach.*

BOTTOM *Bulldozers can clear fuels from a fireline very efficiently. They can also be used to construct escape routes, safety zones, roads and in mopping-up operations.*

BELOW RIGHT *A firefighter lights a backfire. A fire is set from a control line like a road toward the fire's edge. Drafts created by the main fire will pull the fires together. The back fire eliminates fuel and reinforces the fireline.*

On May 14, ten days after the initial blaze, Los Alamos residents were able to board buses and tour the town to see their charred homes. People were emotional but quiet as they viewed the ruins. Angry homeowners demanded explanations for the fire.

Families whose homes were destroyed were offered three-bedroom mobile homes for up to 18 months. Congress earmarked $660 million for wildfire damage, which included $500 million for home and business owners as well as $138 million for the laboratory.

Alpine Train, Kaprun, Austria

DATE · NOVEMBER 11, 2000
DAMAGE · 155 DEAD
CAUSE · UNKNOWN

Austria's worst alpine disaster occurred on a beautiful fall morning when 180 skiers and snowboarders boarded the funicular train pulled by cables up the popular Kitzsteinhorn mountain southwest of Salzburg. Many of the passengers were children and teenagers headed to a snowboarding competition on the glacier.

The blaze erupted at 9:30 a.m. when the train had traveled about 600 yards into the tunnel. Austrian Gerhard Hanetseder, one of the few survivors, said smoke began to fill the lift cabin soon after

The Railroad Before the Fire

Built in 1974, the railway links the resort of Kaprun with the glacier on the north side of the Kitzsteinhorn, 3,202 meters high. The mountain contains many high altitude ski runs and a restaurant at the top. The railway had been modernized in 1994.

Harold Wimmer, an official, said, "The fire took hold at the back of the funicular, in the second carriage…those who were above the fire had no choice but to climb upwards." Those who headed up the tunnel were climbing hundreds of steps in the dark on a ladder that was red hot. The tunnel acted like a huge chimney sending poisonous smoke and deadly fumes upwards.

BELOW *Officials, inspecting the melted carriage of the train, look for clues that would explain what started the fire.*

the train left the station at the base of the mountain. Flames were visible after the train entered the 2-mile long tunnel. "Then panic spread, we tried desperately to get the doors open…in the meantime, the entire cabin was on fire," said Hanetseder. Passengers used their ski poles and boots to smash the windows.

The operator had reported the fire and the train was stopped but the doors were not opened and the lights went out. Survivors spoke of panic and screaming as passengers, hampered by their heavy ski boots, attempted to climb over sports equipment.

The fire was so hot that the aluminum body of the car was burnt down to the chassis. The temperature reached 1,000 degrees and completely incinerated many of the victims. The fire continued to burn fiercely for two days.

Rescue workers had to wait for toxic fumes to dissipate before they could enter the tunnel. They found charred remains that could not be identified because the noses, ears, lips and eyes had all been burned away. The car had completely melted, forcing rescuers to cut out each victim. The task was very slow because only six workers were allowed in the tunnel at once and it took about 45 minutes to extract each body. Working in 90–minute shifts, rescuers had to climb down 2,000 steps from the middle station and load the remains onto a service unit which then took them to the top. The job was made more difficult because there was a danger that the roof of the tunnel might collapse and because there were high winds.

The bodies retrieved were flown to Salzburg for DNA identification. By the 16th the last body was cut from the wreckage. Pathologists and forensic scientists said it would take several weeks to identify the victims. Three people were killed in the station at the top of the tunnel when poisonous smoke overwhelmed them, as the driver of the empty second train was descending through the tunnel in tandem with the burning car.

Among the dead were ten Japanese, 8 U.S. citizens including military personnel and their families based in Germany, 2 Dutch, 4 Slovenes, 1 Czech, 1 Briton and 92 Austrians.

ABOVE *Rescue workers survey the track that leads into the tunnel where the disaster struck.*

Freestyle skiing champion Sandra Schmitt, 19, who was an Olympic hopeful, was killed with her parents.

One Vienna newspaper called it "the nation's worst catastrophe since 1945."

Experts believe that the synthetic ski clothes, lacquered plastic snowboards, skis, boots and upholstery contributed to the poisonous smoke. "The outfit and snowboard of just a single person would have the same burning capacity as 30 liters of heating oil," a fire expert said.

An oily residue was found on the tracks on the 14th and investigators were examining whether it contributed to the fire. Investigators said it could be months before a cause was known.

"The idea that a fire could break out in the tunnel was considered to be something unimaginable."
—Peter Praauer, Commercial Director, Gletscher Bahnen AG train firm

The Reaction

The fire prompted calls for sprinkler systems in tunnels and fire extinguishers in the train cars. Swiss inspectors examined all the ski lift systems following the fire and the resort reopened its slopes a week later.

Photo Credits

23rd Street Fire As It Happened, Frank Cull: p.180, p.181, p.182

AKG Photo: p.36, p.37

AP/Wide World Photos: p.127, p.128, p.129, p.130, p.133, p.134 (bottom left), p.142, p.152, p.153, p.155, p.157, p.156, p.158, p.159, p.160, p.162, p.163, p.164, p.165, p.167, pp.168–169, p.170, p.171, pp.172–173, p.174, p.176, p.177, p.190, p.196, p.197, p.198, p.199, p.202, p.203, p.204, p.205, p.206, p.211, p.213, p.216, p.217, p.218, p.222, p.223, p.224, p.225, p.226 (top and bottom), p.228, p.235, p.236, p.239, p.243

Archive Photos: p.13 (top), p.16, p.27, p.53, p.55, p.62 (bottom), p.104, p.113, p.135, p.137, p.149, pp.186–187, p.214, p.215, p.215, p.220, p.221, p.238, p.246, p.247

Atlantic City Historical Society: p.97

Boston Pictorial Archive: p.66, p.67

Bridgman Art Library International Ltd: p.13 (bottom), p.28, pp.30–31

Brown Brothers: p.94, p.105, p.119 (bottom)

California State Library: pp.42–43

Chicago Historical Society: p.63, p.86 (top and bottom), p.87, p.100 (left), p.175

Corbis-Bettmann/UPI: p.12, p.26, p.58 (left), p.59, p.71, p.107 (bottom right), p.112, pp.124-125, p.126 (top), p.131, p.139, pp.140–141, p.143, pp.146–147,

p.166, pp.178–179, p.183, p.191, pp.192–193, p.194, p.195, p.209, p.210, p.212, p.219, p.244

Der Grofe Brand und der Wiederaufbau von Hamburg, Julius Faulwasser: p.37 (top left and right)

Duke University: p.108 (top)

FDR Library: p.145 (top and bottom)

Gamma: p.207, p.230, p.231, p.232, p.233, p.237

Gibbes Museum of Art/CAA: pp.14–15

Historical Society of Western Pennsylvania: p.38, p.39

Historical Society of Berks County: p.109, p.110

History of Milwaukee, John G. Gregory: p.44, p.45 (top left and right)

Illustrated London News Picture Library: pp.40–41, p.72, p.78, p.79, p.80, p. 90, p.91

Hulton Getty/Liaison Agency: p.138, p.144

Library of Congress: p.52, p.61, p.64, pp.98–99, p.103, pp.106–107 (top), p.108 (bottom), p.114, p.115 (top and bottom), pp.117–118 (top), p.118 (top left and right), p.134 (bottom right)

London's Burning, John Bedford: p.11

Mary Evans Picture Library: p.23, p.29, p77, p.132, pp.136–137

Maryland Historical Society: p.101, p.102

Milwaukee County Historical Society: p.73, p.74

Minnesota Historical Society: p.84, p.85, p.88

Museum of History and Industry: p.81, p.82, p.83 (top and bottom)

Museum of the City of New York: p.19 (top), pp.24-25, p.32, p.46, p.48 (top), p.89 (right)

New Bern-Craven County Photographic Archive: p.121, pp.122–123 (all)

New York Public Library: p.17, p.18, p.19 (bottom), p.20, p.33, p.35, p.45 (bottom), p.47, p.48 (bottom), p.50 (bottom), p.54, p.56, p.57, p.58 (right), p.62 (top), p.68, p.69, p.70 (top and bottom), p.93 (left and right), p.116 (bottom), p.119 (top), p.120

Northern Illinois Flight Center: p.208 (top and bottom), p.210 (right top and bottom), p.218 (right), p.227, p.234 (all), p.245 (all)

Rensselaer County Historical Society: p.49, p.50 (top), p.51 (top and bottom)

Rosenberg Library, Galveston, Texas: p.75, p.76 (top and bottom)

The Gazetteer of Bombay City and Island, Volume II, B. G. Kunte: p.21, p.22

U.S. Army Signal Corps: p.150, p.151

Selected Bibliography

Arensberg, Charles F.C. "The Pittsburgh Fire of April 10, 1845." *The Western Pennsylvania Historical Magazine*, Mar.-June, 1945, v.28, n.1–2, p.11–19.

Austin, H. Russell. *The Milwaukee Story: The Making of an American City*. Milwaukee: Journal Co., 1946.

Baldwin, Leland D. *Pittsburgh: The Story of a City*. Pittsburgh: University of Pittsburgh Press, 1937.

Beasley, Ellen. *The Alleys and Back Buildings of Galveston: An Architectural and Social History*. Houston: Rice University Press, 1996.

Bedford, John. *London's Burning*. London: Abelard-Schuman, 1966.

Bell, Walter George. *The Great Fire of London in 1666*. London: John Lane, 1920.

Benzaquin, Paul. *Holocaust!* New York: Henry Holt and Co., 1959.

Bond, Horatio. *Fire and the Air War*. Boston: National Fire Protection Association, 1946.

Bryant, William O. *Cahaba Prison and the Sultana Disaster*. Tuscaloosa: University of Alabama, 1990.

Burg, David F. *Chicago's White City of 1893*. Lexington: University Press of Kentucky, 1976.

Carter, Samuel, III. *The Siege of Atlanta, 1864*. New York: St. Martin's Press, 1973.

Conway, W. Fred. *Firefighting Lore: Strange but True Stories from Firefighting History*. New Albany, IN: FBH Publishers, 1996.

Cowan, David and John Kuenster. *To Sleep with the Angels: The Story of a Fire*. Chicago: Ivan R. Dee, 1996.

Cudahy, Brian J. *Around Manhattan Island and Other Maritime Tales of New York*. New York: Fordham University Press, 1907.

Firefighters Handbook: Essentials of Firefighting and Emergency Response. Albany: Delmar Publishers, 2000.

Frazer, Sir James George. *Myths of the Origin of Fire*. New York: Hacker Art Books, 1974.

Freedman, Jean R. *Whistling in the Dark: Memory and Culture in Wartime London*. Lexington: University of Kentucky Press, 1999.

Gazetteer of Bombay City and Island, Vol. II. Pune, India: Government Photozinco Press, 1978.

Goldstein, Donald M., Katherine V. Dillon, J. and Michael Wenger. *Rain of Ruin: A Photographic History of Hiroshima and Nagasaki*. New York: Brasseys, 1995.

Gordon, Leonard. *A City in Racial Crisis: The Case of Detroit Pre- and Post- the 1967 Riot*. Dubuque, Iowa: Wm. C. Brown Co., 1971.

Hazen, Margaret Hindle and Robert M. Hazen. *Keepers of the Flame: The Role of Fire in American Culture, 1775–1925*. Princeton: Princeton University Press, 1992.

Hearsey, John E. N. *London and the Great Fire*. London: John Murray, 1965.

Hersey, John. *Hiroshima*. New York: Alfred A. Knopf, 1985.

Heys, Sam and Allen B. Goodwin. *The Winecoff Fire: The Untold Story of America's Deadliest Hotel Fire*. Atlanta: Longstreet Press, 1993.

Kern, Edward (Sonny). *The Collinwood School Fire of 1908*. Private printing, 1993.

Keyes, Edward. *Cocoanut Grove*. New York: Atheneum, 1984.

Latham, Robert and William Matthews, eds. *The Diary of Samuel Pepys*, vol. VII. -1666. Berkeley: University of California Press, 1972.

Leighton, Isabel, ed. *The Aspirin Age 1919–1941*. New York: Simon and Schuster, 1949.

Maddox, Robert James. *Weapons for Victory: The Hiroshima Decision Fifty Years Later*. Columbia, MO: University of Missouri Press, 1995.

Mommsen, Theodor. *A History of Rome Under the Emperors*. London: Routledge, 1996.

Muccigrosso, Robert. *Celebrating the New World: Chicago's Columbian Exposition of 1893.* Chicago: Ivan R. Dee, 1993.

Offrey, Charles, et al. *Normandie: Queen of the Sea.* New York: Vendome Press, 1985.

Olivier, Daria. *The Burning of Moscow.* London: George Allen & Unwin Ltd., 1966.

O'Nan, Stewart. *The Circus Fire: A True Story.* New York: Doubleday, 2000.

Parramore, Thomas C. *Norfolk: The First Four Centuries.* Charlottesville: University Press of Virginia, 1994.

Patrick, Rembert W. *The Fall of Richmond.* Baton Rouge: Louisiana State University Press, 1960.

Pernin, Reverend Peter. *The Great Peshtigo Fire: An Eyewitness Account.* 2nd ed. Madison: State Historical Society of Wisconsin, 1999.

Potter, Jerry O. *The Sultana Tragedy: America's Greatest Maritime Disaster.* Gretna, Pelican Publishing, 1992.

Pyne, Stephen J. *Fire in America: A Cultural History of Wildland and Rural Fire.* Seattle: University of Washington Press, 1982.

Pyne, Stephen J. *Vestal Fire: An Environmental History, Told Through Fire, of Europe and Europe's Encounter with the World.* Seattle: University of Washington Press, 1997.

Ray, John. *The Night Blitz 1940–1941.* London: Arms and Armour Press, 1996.

Rosen, Christine Meissner. *The Limits of Power: Great Fires and the Process of City Growth in America.* Cambridge: Cambridge University Press, 1986.

Teie, William C. *Firefighter's Handbook on Wildland Firefighting: Strategy, Tactics and Safety.* Rescue, CA: Deer Valley Press, 1994.

Tully, Andrew. *When They Burned the White House.* New York: Simon and Schuster, 1961.

Walker, J. Samuel. *Prompt and Utter Destruction: Truman and the Use of Atomic Bombs Against Japan.* Chapel Hill: University of North Carolina Press, 1997.

Wells, Robert W. *Fire at Peshtigo.* Englewood Cliffs: Prentice-Hall, 1968.

Index